Richard Socher

Automatic Extension of Semantic Lexicons with a Bootstrapping Algorithm

Richard Socher

Automatic Extension of Semantic Lexicons with a Bootstrapping Algorithm

Using Corpora to Learn Semantic Features

VDM Verlag Dr. Müller

Imprint

Bibliographic information by the German National Library: The German National Library lists this publication at the German National Bibliography; detailed bibliographic information is available on the Internet at http://dnb.d-nb.de.

Any brand names and product names mentioned in this book are subject to trademark, brand or patent protection and are trademarks or registered trademarks of their respective holders. The use of brand names, product names, common names, trade names, product descriptions etc. even without a particular marking in this works is in no way to be construed to mean that such names may be regarded as unrestricted in respect of trademark and brand protection legislation and could thus be used by anyone.

Cover image: www.purestockx.com

Publisher:
VDM Verlag Dr. Müller Aktiengesellschaft & Co. KG, Dudweiler Landstr. 125 a, 66123 Saarbrücken, Germany,
Phone +49 681 9100-698, Fax +49 681 9100-988,
Email: info@vdm-verlag.de

Copyright © 2008 VDM Verlag Dr. Müller Aktiengesellschaft & Co. KG and licensors
All rights reserved. Saarbrücken 2008

Produced in USA and UK by:
Lightning Source Inc., La Vergne, Tennessee, USA
Lightning Source UK Ltd., Milton Keynes, UK
BookSurge LLC, 5341 Dorchester Road, Suite 16, North Charleston, SC 29418, USA

ISBN: 978-3-639-03137-9

Abstract

This work investigates and improves a bootstrapping approach which permits to extend high quality lexical resources with the help of large corpora. The emphasis lies on the extraction of lexical-semantic information and word meaning, which are fundamental components for advanced applications such as semantic parsing, information retrieval or summarizing textual information.

The approach is based on co-occurrences of verbs with nouns in a specific context such as object, subject or certain theta-roles. The experiments use a large parsed corpus and are compared to past investigations with adjectives and nouns in order to find out whether adjective modifiers or certain verb - noun relations are more suitable for classifying nouns with respects to their semantic characteristics. The algorithm starts with several seed words whose characteristics are known and which stand in certain relations to the respective verb or adjective in the sentence. Other unknown nouns that co-occur in the same context then inherit some of the characteristics of these seed words.

The aim is to find the most effective relations to enhance semantic resources for nouns in general and apply the findings to the German lexicon HaGenLex automatically and weakly supervised. The findings of this work will help to extend already existing lexicons. This is necessary since there are still no sufficiently large semantic lexicons for German.

The first chapter outlines computer linguistics and corpus linguistics and explains the semantic structure that is used in the lexicon. Furthermore basics of bootstrapping and other similar approaches are provided to better understand the scientific context of this work and to show the general applicability of such an approach.

In chapter 2 the pre-processing steps are presented and the algorithm is explained by theory and examples. Different parameters that can result in major changes of the results are shown.

Chapter 3 describes the experiments in great detail. While experiments with adjectives have already been done and are compared to new experiments, the extension to verbal relations such as subject, object and theta-roles has hitherto not been examined for German. By means of extensive experiments, effective relations for bootstrapping are discovered and optimal new parameter combinations are found. The chapter ends with the combination of the three main relations, which outperforms separately obtained solutions and increases precision significantly.

In the last chapter an outlook with suggestions for further improvements and extensions is given and an absolutely novel approach which combines genetic algorithms and bootstrapping is outlined.

Acknowledgements

Thanks to Prof. Gerhard Heyer and Chris Biemann for their support and professional advice. Furthermore I would like to thank Rainer Osswald and Sven Hartrumpf from the Fernuniversität Hagen for their support with the lexicon, corpus and extraction tools. Foremost, I would like to thank my parents for their enduring help before and during my studies.

Contents

1. **Introduction**   1
   - 1.1. Computer Linguistics 2
     - 1.1.1. Definition 2
     - 1.1.2. Brief History 3
     - 1.1.3. Application Domains 3
   - 1.2. Corpus Linguistics 4
     - 1.2.1. Definition 4
     - 1.2.2. Brief History 4
     - 1.2.3. Corpus Features 4
     - 1.2.4. Criticism 5
     - 1.2.5. Summary 5
   - 1.3. General Definition of a Lexicon 5
   - 1.4. Theta Roles 6
     - 1.4.1. Definition 6
     - 1.4.2. Different Views 6
   - 1.5. The HaGenLex Lexicon 7
     - 1.5.1. Overview of the MultiNet Paradigm 7
     - 1.5.2. Ontological Sorts 8
     - 1.5.3. Features 8
     - 1.5.4. Semantic Sorts 9
     - 1.5.5. Theta-Roles and Other Lexical Semantic Relations 10
   - 1.6. Other Lexical and Semantic Resources for German 11
   - 1.7. Corpus Based Acquisition for German Lexicons 12

2. **Bootstrapping Algorithm**   14
   - 2.1. Other bootstrapping methods for the acquisition of semantic lexicons . . 14
   - 2.2. Pre-processing Steps for Input Data 16
     - 2.2.1. Parsing the Corpus 17
     - 2.2.2. Extracting Context Relations 20
     - 2.2.3. Extracting Statistical Significant Co-occurrences 21
     - 2.2.4. Seed Words - Nouns in HaGenLex 24
     - 2.2.5. Bipartite Graph with Bootword-Noun Pairs 26
   - 2.3. The Algorithm 27
     - 2.3.1. Classification 27

	2.3.2. Principle	27
	2.3.3. Example	29
2.4.	Assignment of Complex Semantic Sorts	31
	2.4.1. Summary	32

3. Experiments, Results and Interpretation 33

- 3.1. Corpus, Extracted Co-occurrences and Noun Database 34
  - 3.1.1. Results from Wikipedia and CLEF Corpus 34
  - 3.1.2. Possible Drawbacks of the Wikipedia Corpus 35
  - 3.1.3. Corpus: Projekt Deutscher Wortschatz 36
- 3.2. Experiment Settings 36
  - 3.2.1. Experiment Measures 36
  - 3.2.2. Parameters for First Experiments with the Three Major Groups . 37
- 3.3. Adjective - Noun 39
  - 3.3.1. Features 39
  - 3.3.2. Ontological Sorts 41
  - 3.3.3. Combination to Semantic Sorts 41
- 3.4. Verb - Noun(Subject) 42
  - 3.4.1. Features 43
  - 3.4.2. Ontological Sorts 43
  - 3.4.3. Combination to Semantic Sorts 44
- 3.5. Verb - Noun(Object) 45
  - 3.5.1. Features and Ontological Sorts 46
  - 3.5.2. Combination to Semantic Sorts 46
- 3.6. Verb - Noun(Theta-Role) 48
  - 3.6.1. Features and Ontological Sorts 49
  - 3.6.2. Semantic Sorts 51
- 3.7. Comparison of Adjective-Noun Relations and Verbal Contexts 52
  - 3.7.1. F-score of the Three Main Groups 52
  - 3.7.2. Verification and Conclusion 53
  - 3.7.3. Combination to Semantic Sorts 54
  - 3.7.4. Doubt-Measure 54
  - 3.7.5. Theta-Roles 55
- 3.8. Filtering for all Significant Co-occurrences 56
- 3.9. Effects of Smoothing 56
- 3.10. Effects of Parameter Changes: minBoot and maxClass 57
- 3.11. Combination of Modifying and Verbal Relations 59
  - 3.11.1. Binary Features and Ontological Sorts 59
  - 3.11.2. Composing Semantic Sorts with Combined Results 61
- 3.12. Error Analysis 63
- 3.13. Summary 64

Contents

4. Improvements and a Novel Approach   **66**
    4.1. Combination of Characteristics to Complex Semantic Sorts   66
    4.2. Selection Procedure between Runs   67
    4.3. Using the Doubt Measure   67
    4.4. Genetic Bootstrapping   68
        4.4.1. Genetic Algorithms   68
        4.4.2. Genetic Algorithms in NLP   69
        4.4.3. The Novel Approach   69

5. Conclusion   **72**

Glossary   **74**

A. HaGenLex Tables   **77**

B. Bootstrapping Results   **79**

Bibliography   **83**

List of Figures

1.1. Features of the semantic sorts *con-potag* ⊃ *animate-object* ⊃ *animal-object* 9
2.1. Scheme with pre-processing steps for bootstrapping 17
2.2. A netarx query for the theta-role *method* 21
2.3. Example with extracted adjective-noun pairs, showing numbers for each possible polysemous meaning 21
2.4. Example of a bipartite graph 27
2.5. Bootstrapping algorithm for assigning semantic attributes to nouns . . . 28
2.6. An excerpt from German verb-object relations 30
2.7. Abstract visualization of an excerpt from German verb-object relations . 30
2.8. Algorithm for combining characteristics to complex semantic sorts 31

3.1. Adjective-noun: features - precision and recall 40
3.2. Adjective-noun: sorts - precision and recall 42
3.3. Verb - noun(subject): features - precision and recall 44
3.4. Verb - noun(subject): sorts - precision and recall 45
3.5. Verb - noun(object): features - precision and recall 47
3.6. Verb - noun(object): sorts - precision and recall 47
3.7. F-score of all relations 49
3.8. Adjective, subject and object f-scores, based on bias 52
3.9. Example doubt values 55
3.10. Different results based on filtering for significant pairs. 57

4.1. Algorithm for genetic bootstrapping 71

List of Tables

1.1.	HaGenLex theta-roles	10
2.1.	Table for finding the significance level based on log-likelihood ratio	24
2.2.	Small excerpt of the 50 possible semantic sorts	26
2.3.	Characteristics of the semantic sort *human-object*	26
3.1.	Pre-processing: number of extracted pairs from Wikipedia and CLEF corpus	35
3.2.	Adjectives: features, distribution, bias and bootstrapping results	40
3.3.	Adjectives: sorts, distribution, bias and bootstrapping results	41
3.4.	Subjects: features, distribution, bias and bootstrapping results	43
3.5.	Subjects: sorts, distribution, bias and bootstrapping results	44
3.6.	Objects, mean values: features and sorts, distribution, bias and bootstrapping results	46
3.7.	Objects: Combination to semantic sorts - best results	48
3.8.	Correlation between number of extracted pairs and average F-score	49
3.9.	Theta-roles and some single characteristics with above 90% precision	50
3.10.	Theta-Role *aff*: High precision values	51
3.11.	Projekt Deutscher Wortschatz: Bootstrapping precision for adjective, object and subject relations	53
3.12.	Results with different parameter values of maxClass and minBoot	58
3.13.	Combining results of the three main relations, parameter combination: 2-1	60
3.14.	Combining results of the three main relations with parameters in single runs: maxClass=2 and minBoot=1	61
3.15.	Semantic sorts composed from the combination of three equal values from the three main relations, which were bootstrapped separately with the parameters 2-5	62
A.1.	Semantic features ([Sch98])	77
A.2.	Hierarchy of ontological sorts ([IHIO03])	78
B.1.	Adjectives: Results of combination to semantic sorts	79
B.2.	Subjects: Results of combination to semantic sorts	79
B.3.	Objects: features, distribution, bias and bootstrapping results	80
B.4.	Objects: sorts, distribution, bias and bootstrapping results	80

List of Tables

B.5. Objects: Correct classifications of smaller (positive) classes 80
B.6. Objects: Results of combination to semantic sorts 81
B.7. All bootstrapping results for theta-role *agt* 81
B.8. Combining results of the three main relations, parameter combination: 2-5 82

Chapter 1.

Introduction

> A Word that breathes distinctly
> Has not the power to die
> Cohesive as the Spirit
> It may expire if He -
> 'Made Flesh and dwelt among us'
> Could condescension be
> Like this consent of Language
> This loved Philology.
> - *Emily Dickinson* -

Where is the information I need in this huge pile of data? This question will be asked more and more in our information society and the only way to solve it on a large scale is to process the data automatically. Since most often this data is unstructured and available only in natural language, we need to understand the inner structure of natural languages and employ the tools of computer science to effectively extract the information we need.

For this task to be successful a good lexicon is needed. Many fields of advanced natural language processing, such as information retrieval (IR), word sense disambiguation or semantic web applications are based on lexical information. Since the manual creation of high quality lexical resources is very expensive and time consuming, there is a need for automatic or semi-automatic tools to create these. As of now no large high-quality lexical-semantic resources are available for German.

This work investigates and extends a bootstrapping approach, which permits to extend high quality lexical resources with the help of very large corpora. The emphasis lies on the extraction of lexical-semantic information. The idea is to start with a given number of known nouns that co-occur with a certain type of bootword such as adjectives or verbs in a specific context. Based on these co-occurrences in a large corpus, each bootword gets assigned a hypothesis for certain noun characteristics. After this training phase, unknown nouns with the same context can then inherit values for each characteristic from the bootword and can be added to the lexicon.

Chapter 1. Introduction

This method shall be used to systematically expand a verified syntactico-semantic lexicon with 11,100 nouns, based on a corpus of about 10 million semantically annotated sentences. The following relations are extracted for bootstrapping:

- adjective - noun
- verb - noun (subject)
- verb - noun (object)
- verb - noun (in specific theta role)

For general explanations the relations between nouns and adjectives or the relations between object-nouns and subject-nouns respectively are abbreviated with *adjectives*, *objects* and *subjects*.

While the method has already been successfully used for co-occurrences of adjectives and nouns ([BO05]), it has hitherto not been extended to verbs and their nouns in different functional positions.

The first chapter gives an introduction to the field of corpus linguistics and computer linguistics and outlines the structure of the lexicon HaGenLex, which is based on multilayered extended semantic networks.

The following chapter describes the bootstrapping algorithm itself together with the anticipated behavior and the necessary pre-processing steps in detail. Afterwards the results are analyzed and single characteristics are merged to more complex classes.

Possible improvements and extensions of the presented method are explained in the last chapter.

1.1. Computer Linguistics

1.1.1. Definition

Computer linguistics[1] is an interdisciplinary area of research between linguistics and computer science. Its goals are to discover and complete linguistically relevant knowledge and simulate language processes. It strives for explicit or procedural description of these processes and their application in practical software. Being embedded in an extensive framework linguistic and computer linguistic findings are not necessarily distinct. Differences are usually found in the methods of both.

---

[1] Also called computational linguistics or linguistic computer science and abbreviated with CL.

Chapter 1. Introduction

1.1.2. Brief History

The term *computational linguistics* as the name of a new field of research was introduced in the beginning of the sixties of the 20th century by David G. Hays. But even before, in the fifties the first pioneers started to work in the field of machine translation. At that time problems with calculations seemed bigger than those of the theory and structure behind natural languages. At the end of the sixties the focus changed from the use of computers to solve problems to the overcoming of problems with the help of computers. Language theories gained more importance and as a result of that Chomsky's transformation grammar and syntax in general. In the following years three areas had a significant impact on computer linguistics:

- machine translation that focused on syntax

- information retrieval which deals with semantics, formal representations of world knowledge and inference rules

- and question-answer systems that combined both with aspects of communication.

The third became especially popular in the seventies. Lexicons, syntax and semantic interpretation started to be treated more modular. Martin Kay's unification based grammar models were an important milestone since they allow to define an incomplete set of descriptive rules that can be extended incrementally over time.([BLP89])

1.1.3. Application Domains

In contrast to linguistics, which tries to describe natural language under different aspects such as phonetics, phonology, morphology, syntax, semantics and pragmatics, computer linguists also try to reproduce language and develop applications that implement these theories effectively. Furthermore new fields have been created by CL, such as:

- information retrieval,

- document classification,

- human-machine interaction,

- automatic grammar and orthography corrections and

- text-to-speech systems.

Specific areas such as corpus linguistics experienced a revival due to their interaction with computers and the possibilities they provide.

Chapter 1. Introduction

1.2. Corpus Linguistics

1.2.1. Definition

Corpus linguistics is a sub domain of computer linguistics and thus both share the same goals. The main difference lies in the statistical methods of the former which are mostly quantitative linguistic analyses of real world text, so called corpora. A corpus is a text in written or spoken form. In this setting it is available in an electronic form.

1.2.2. Brief History

The first big corpus was created by about 5000 analysts under the supervision of Kaeding in 1897. They created an impressive corpus for their time of about 11 million words in German, mainly to collate frequency distributions of letters and sequences of letters. During that time studies of language acquisition and child language became popular. They were based on dictionaries that parents recorded about their children's progress in learning. One of the goals was to find norms of development for early years of language learning and later also for foreign language studies.

When Avram Noam Chomsky published his book *Syntactic Structures* in 1957 it had a significant impact on corpus linguistics and the predominating empiricism and marked the end of the early corpus linguistics. The fifties were the time of the cognitive revolution whose main goal was to learn about human mental processes. It was in this when research in artificial intelligence and computer science gained importance. The developed linguistic and computational methods were mostly rule based and focused on speaker's competence, that is their innate international knowledge of language. Chomsky hold the opinion that only by studying competence and its finite set of rules one can understand the inner structure of language. Corpora on the other hand were only - what he called - *performance*, i.e. external and sometimes poor evidence of this competence. That is why corpus linguistics were of no big importance during these years. Its revival came with the compilation of the Brown Corpus of Standard American English by Kucera and Francis ([KF64]) and their classic work 'Computational Analysis of Present-Day American English' (Brown Corpus) [KF67]. As computing power and capacity increased greatly during the eighties corpus linguistics became a growing field of research and is regarded as an integral part for language studies today.

1.2.3. Corpus Features

For the framework of today's corpus linguistics the following extended list of corpus features is of importance:

Chapter 1. Introduction

- encoding and annotation, the format in which the corpus is saved and linguistic information such as part-of- speech tags (POS tags), person, number, etc.
- finite size, number of saved characters, tokens and sentences
- technical means to query the corpus,
- representativeness, i.e.: *Is the corpus only based on text from the same domain? Does the distribution of words in the corpus reflect their distribution in natural language?*
- availability to other researchers.
- linguistic meta information, like date, author etc.

1.2.4. Criticism

Statistical methods that are used in corpus linguistics have often been criticized by linguists following rule based methods only. Some of these objections are based on the infinity of language, which necessarily results in skewed corpora. This is comprehensible criticism and has been taken into account by trying to construct representative corpora which is becoming easier with the upcoming of very powerful computer systems. The fact that introspective judgements are unobservable and unlike a corpus not scientifically verifiable is often ignored. Of course there are some cases where introspective thoughts provide better solutions, but tasks like frequency counts, verification of hypotheses from grammatical theories or representative quantification of grammar rules are only possible with corpora.

1.2.5. Summary

In summary, rule based approaches are more precise than statistical methods but the latter scale better and can process more raw data. While the error rate is somehow higher, more problem classes can be treated, which would require more and specific rules for each class ([KYCC96]). This work combines the two in order to gain high recall as well as high precision.

1.3. General Definition of a Lexicon

A lexicon is a database that contains orthographic words and certain corresponding information for each word. Depending on the type of lexicon these can be related to the: morphology, syntax or semantics of the word. One word entry can have different part of speech (POS) tags, the word *intimate* for example can be an adjective or a verb. Depending on this POS tag, inflectional information can be different. One syntactically

Chapter 1. Introduction

specified word can again represent different semantic concepts. The word *school*, which is a noun could represent an institution or a building for example.

Besides these three domains, a lexicon can also save information extracted from corpora, such as the frequency of an orthographical word, or of its word senses. Huge semantic lexicons are a fundamental basic for many advanced fields in computer linguistics.

1.4. Theta Roles

1.4.1. Definition

Theta-roles (also called, thematic roles, θ-roles or cognitive roles) are conceptual role relations between a verb and its nominal arguments. Examples are in square brackets: see-[agent]-monkey, or brake-[patient]-window. Verb valency lists capture restrictions about the possible complements of a verb. Other theta-roles might be adjuncts. These are not optional and thus not in the valency list.

1.4.2. Different Views

Numerous theoretical accounts exist to model this phenomenon. Only three different views shall be outlined. Gruber for example defines an *agent, theme, patient, source* and *goal* ([Gru65]). In the example sentence: 'I gave my thesis to my supervisor.' 'I' is the agent, 'my thesis' is the patient (or theme) and 'my supervisor' is the goal. Jackendoff uses Gruber's empirically based inventory of roles as a basis, adds the *location* role and combines these with conceptual semantics ([Jac87]).

Fillmore ([Fil68]) introduces the concept of deep cases which are very similar to theta-roles and define semantic functions of verb complements. He also uses the role of an agent and object (theme), but also has the roles of instrumental, dative (an animate entity affected by an action), factitive (result), locative (place where the action of the verb takes place) and the neutral object. His deep cases can be semantically realized as subject or object, as in all these theories.

Dowty ([Dow91])states that there is no clear understanding of theta-roles and their necessary level of granularity. He then proposes the possibility to add a role to each verb. Instead of an agent in the valncy list of the verb 'hit', there would be role of a 'hitter', and for 'play' a 'player', etc. However he rejects this idea, because it lacks any syntactic or semantic generalization. For Dowty a theta role is 'a set of entailments of a group of predicates with respect to one of the arguments of each'. For Dowty there are besides the traditional theta-roles of agent, experiencer, instrument and theme also prototype theta-roles. These are proto-agent and proto-theme which both cluster concepts and are characterized by a set of verbal entailments. The former is basically an

Chapter 1. Introduction

active participant in the course of actio, while the latter is undergoing a change of state or is in some way causally effected by another participant. Dowty's ideas are picked up by FrameNet which is outlined in section 1.7.

For a list of theta-roles used in this work and in the MultiNet paradigm, see section 1.5.5

1.5. The HaGenLex Lexicon

Hagen German Lexicon (HaGenLex) is a domain independent computer lexicon for German morphology, syntax and semantics based on the *Multilayered extended semantic networks* (MultiNet) paradigm([Hel01]). It was designed to support various natural language processing (NLP) applications such as question answering systems and IR and provides the necessary lexical information for the bootstrapping algorithm.

HaGenLex is independent from any grammar theory such as head driven phrase structure grammar (HPSG, see [PS94]) with which it shares only similar feature-value structures. It contains about 9200 nouns, 6500 verbs and 3000 adjectives, all manually added and annotated with morphological, syntactic and semantic information. Word meanings are systematically linked to those in the GermaNet database (see [KW01]).

Entries in HaGenLex are classified by ontological sorts. Nouns, restrictions of verb valency lists and adjuncts are also semantically defined by 16 binary features, such as HUMAN, MENTAL or ANIMATE. These two are combined into semantic classes.

The following sections explain the lexical and semantic structure of HaGenLex and its framework MultiNet. The last section explains the way theta-roles are treated inside this framework.

1.5.1. Overview of the MultiNet Paradigm

Multilayered extended semantic networks ([Hel01]) are both a knowledge representation paradigm and a language for meaning representation of natural language expressions. They are used to describe all entries in the lexicon, are the output of the semantic analysis of the corpus and can be used for knowledge modelling and inferences in question-answer systems. A good knowledge of this framework is indispensable for understanding HaGenLex.

MultiNet is neither a model theoretical extensional nor a procedural theory. Instead it is based on the connection between cognitive concepts and expressions. As such it is a lexical structural semantic theory following the thoughts of Wittgenstein. The meaning of expressions and semantic primitives is defined by interrelations and connections of expressions. In this process concepts are created. These are mental pictures of real or imagined entities. In MultiNet there is a bijective map between lexemes and lexicalized concepts.

These concepts are represented in semantic networks, which are labelled directed hypergraphs. The nodes represent concepts, and the directed edges between the nodes represent relations and functions. These establish semantic connections between nodes.

MultiNet also provides an inner structure for nodes, so called layers. Through them each concept is connected to immanent knowledge which connects the term with its meaning and to knowledge based on the situation or the actual usage of the term. The later encodes genericity and definiteness information. These values indicate, e.g., the truth or reality of an entity or its determination of reference. This information is not being used in the presented experiments.

Unlike other approaches which have unlimited relations, MultiNet has a predefined set of 140 relations and functions which are written as labels on the edges in the hypergraphs ([HHO03]). The relations build a second semantic network at a meta level. On this level they are nodes which are connected with axioms and inference rules. This information can be used to conclude lexicalized knowledge about the relations.

While the MultiNet's structure is important, it is essential to know the way HaGenLex specifies its noun entries. This is covered in sections 1.5.2, 1.5.3 and 1.5.4. The term *characteristic* subsumes these three way of classifying nouns.

1.5.2. Ontological Sorts

The 45 ontological sorts of MultiNet are completely implemented in HaGenLex. They form a tree-shaped hierarchy which distinguishes between objects [o], situations [si], situational descriptors [sd], qualities [ql], quantities [qn], graduators [gr] and formal entities [fe] on the first level. Only 17 of these apply to nouns and are thus used during the experiments. Except nouns for quantity and units of measurement, they are all sub sorts of *object*. Even though these sorts are not independent they are saved as binary values for faster computation.

For a complete list, see appendix table A.2. For the 17 noun sorts see section 2.2.4.

1.5.3. Features

Ontological sorts alone are often not fine-grained enough and difficult to manage for restrictions of verb valency lists. Suppose the lexicon wants to restrict the subject of the verb *love* to all human beings. It would need a restriction that says that the first argument of the verb needs to be an element of the subset of human beings. A semantic parser would then need to start the whole inference mechanism to try to find out if the given entity is indeed in one of the subsets of human beings. This is a rather complicated and time consuming process. That is why a fixed set of 16 binary semantic features is introduced for the semantic classification of concepts and the specification of selectional restrictions for verbs.

They allow effective computation and easier manual testing and verification since the lexical entries are much simpler. Also restrictions for selections in subcategorisation and valency lists can be more subtly differentiated and congruency tests in analyses (and tests) are faster. To test whether a concept satisfies a given restriction is now as simple as a unification of feature vectors. Some of the features for nouns are:

- ANIMAL
- ANIMATE
- ARTIF, for artifacts, things created by humans
- INFO, carrier for information.

For a complete list, see appendix table 1.

These features are only loosely related to one another. Thus they are all saved as binary characteristics and bootstrapped separately.

The term characteristic refers to both ontological sorts and features. Characteristics are subsumed under semantic sorts.

1.5.4. Semantic Sorts

The values of semantic features are not independent of each other. For example HUMAN entities are necessarily ANIMATE:

$$[HUMAN+] \Rightarrow [ANIMATE+]$$

To reflect these dependencies, 50 predefined sets of semantically coherent feature and sort combinations are defined. They are called semantic sorts.

The following figure shows three semantic sorts that inherit their features from left to right: $con\text{-}potag \supset animate\text{-}object \supset animal\text{-}object$:

$$\begin{bmatrix} \text{con-} & \text{potag} \\ SORT & d \\ AXIAL & + \\ GEOGR & - \\ INFO & - \\ POTAG & + \end{bmatrix} \begin{bmatrix} \textbf{animate-} & \textbf{object} \\ ANIMATE & + \\ ARTIF & - \\ INSTRU & - \end{bmatrix} \begin{bmatrix} \textbf{animal-} & \textbf{object} \\ ANIMAL & + \\ HUMAN & - \\ LEGPER & - \\ MOVABLE & + \end{bmatrix}$$

Figure 1.1.: Features of the semantic sorts $con\text{-}potag \supset animate\text{-}object \supset animal\text{-}object$

Chapter 1. Introduction

Notice that semantic sorts are represented by feature-value structures. This is essential for the presented algorithm since each one of these features can be bootstrapped separately.

1.5.5. Theta-Roles and Other Lexical Semantic Relations

The MultiNet formalism uses standard semantic relations such as synonymy, subordination, hyponymy, antonymy or meronymy (part-of). These relations are incorporated into HaGenLex by connecting lexicalized concepts with each other.

The second group of semantic relations are theta-roles. A general introduction of this concept is presented in section 1.4. MultiNet's approach is partly based on Fillmore's theory. It distinguishes between obligatory elements of a verb's argument list (also called valency list or valency frame), named fillers and circumstances, which are optional. In the center of each semantic net of a sentence is the verb. It also defines the sort of the facts of the matter of the sentence (process or state). In general it is abstracted from syntax in order to find an underlying, language independent deep case. Table 1.1 lists all possible theta-roles that can be fillers for some verbs (table based on [Hel01] and [HHO03]). The signature defines what kinds of ontological sorts can fill the position opened by this verb's valency list. In most cases this is a *si*tuation, denoted by a verb which is connected to an *o*bject. Not only verbs can have such a valency list with theta-roles, but also nouns and adjectives. That is why situational objects (*abs*), such as the ones denoted by the nouns *race* or *robbery* are also possible values for the first element of the relations.

Relation	Signature	Short description
AFF	$[si \cup abs] \times [si \cup o]$	Affected object
AGT	$[si \cup abs] \times o$	Agent
AVRT	$[dy \cup ad] \times o$	Averting/Turning away from an object
BENF	$[si \cup abs] \times [o \setminus abs]$	Benefactee
CSTR	$[si \cup abs] \times o$	Causator
EXP	$[si \cup abs] \times o$	Experiencer
INIT	$[dy \cup ad] \times [si \cup o]$	Relation specifying an initial state
INSTR	$[si \cup abs] \times co$	Instrument
MCONT	$[si \cup o] \times [si \cup o]$	Mental content
METH	$[si \cup abs] \times [dy \cup ad \cup io]$	Method
OBJ	$[si \cup o] \times [si \cup o]$	Neutral object
OPPOS	$[si \cup o] \times [si \cup o]$	Entity being opposed by a situation
ORNT	$[si \cup abs] \times o$	Orientation towards something
RSLT	$[si \cup abs] \times [si \cup o]$	Result
SUPPL	$[si \cup abs] \times o$	Supplement

Table 1.1.: HaGenLex theta-roles

Chapter 1. Introduction

Unfortunately not all elements in this list can be used for the bootstrapping method. The problem is that some roles are most often not elements of verb valency lists, but adjuncts. Adjuncts are optional and thus harder to identify by the semantic parser (which could otherwise try to find elements in the valency list of the verb). Sometimes it can only identify a theta role correctly, if the semantic features and ontological sorts are known.

The problem is thus that the bootstrapping process would use a certain theta-role, in order to find its semantic classes, but the parser was only able to identify this theta-role because it already had this information.

Due to this correlation the following theta-roles, which are most often occurring as adjuncts are not examined and only listed for completeness:

- meth
- instr
- init
- suppl

1.6. Other Lexical and Semantic Resources for German

This section gives a brief summary of available lexical-semantic resources for German. This demonstrates the need for automatic approaches to extend these resources and shows that a large high quality semantic lexicon is still not available.

The most important freely available lexicon is certainly GermaNet [KW01], which has been developed between 1996 and 1999 at the University Tübingen. It is based on WordNet ([MBF+90]) and thus deals extensively with lexical relations such as synonymy, hyponymy and hyperonymy. In contrast to the lexicon HaGenLex which is used in this project it does not collect semantic valence lists nor does it try to provide a complete formalism for semantics.

The CISLEX dictionary from the Centrum für Informations- und Sprachverarbeitung (CIS) at the Universität München includes 180,000 lemmata (without proper nouns and compound words), but is only partly semantically annotated ([LMO96]) and not freely available. It includes a semantic classification for simple nouns. A subset of its verbs is annotated with valency lists and sorts. This approach is not sufficient to completely model German semantics and is not suitable for the bootstrapping algorithm since the semantic information is not fine-grained enough.

In 2000 the SIMPLE project finished. It was funded by the European Union and its goal was to add semantic information to twelve morphosyntactic lexicons of European languages.([LBB+00]) The underlying structure was based on an extension of Pustejovsky's ([Pus95]) theory of qualia structures. These are used for structuring the semantic types and for expressing multi-dimensional, orthogonal aspects of word meaning. Unfortunately no access has been granted to other researchers yet.

Finally the Verbmobil project ([WW00]) provides access only to a narrow set of German which is related to appointments, hotel bookings and the like. It is thus not suitable as a starting point for a complete covering of German semantics.

HaGenLex from the Fernuniversität Hagen is a semantic lexicon of great depth and with a quantity that is a good base for automatic bootstrapping methods. It provides a homogeneous description on word, sentence and text level. Further detail is presented in the corresponding section.

1.7. Corpus Based Acquisition for German Lexicons

This section provides an overview of extensions and refinements of lexicons based on statistical methods on corpora. For the creation of huge lexicons corpora are necessary to ensure adequacy and coverage of the entries and to discover new unknown words. Corpora also help to find the most frequent words that subsequently need to be covered in a dictionary that is restricted in size and can be used to extract multi word concepts and idioms ([ML01]). The presented method demonstrates the fact that lexical resources can be successfully created with the help of corpora.

The goal of the 'Saarbrücken Lexical Semantics Acquisition' (SALSA) project is to build a large frame-based semantic lexicon ([EKPP03]). The first step in this project is to manually construct a semantically annotated corpus based on the FrameNet paradigm for English ([FBS02]). In FrameNet lemmas such as verbs or nouns are associated with a semantic frame. A frame is a script-like conceptual structure that describes a particular type of situation, object, or event and its participants. During the second phase, technical means will be explored to automate the annotation process. The following two works demonstrate how syntactic and semantic information can be extracted with the help of statistical methods applied on big newspaper corpora which are annotated with the Stuttgart-Tübingen-Tagset (see [STST95]). In [EH96] statistics on tag sequences are extracted to find subcategorization frames of verbs. Tag sequences are also the means to discover adjectives that modify multi word subject clauses in [HK02].

Unlike these three approaches, which are based mostly on syntactical evidence there are also works that successfully extract semantic information. [iWSR+01] for example finds homogeneous groups of nouns, based on the verbs they co-occur with. So are *meat, banana, bread* in the same group, since they all occur as a direct object of the verb *eat*. The algorithm finds about 50 homogeneous noun groups with a precision of above 93 %. Since some orthographic words can be homonymous and depict different concepts depending on the context, they can be in more than one semantic group. In order to

Chapter 1. Introduction

find optimal parameters the Expectation-Maximization-Algorithm is used on top of a probabilistic parser.

The next section closely examines the lexicon HaGenLex and the framework under which the acquisition of new entries takes place.

Chapter 2.

Bootstrapping Algorithm

The term bootstrapping is used in numerous scientific fields such as biology, law, electronics, statistics and linguistics. In NLP it refers to a process where few available information (often in the form of seed words) is used as a basis to create more information with the help of a corpus. The emphasis in this approach lies on the extraction of lexical-semantic information and word meaning, based on the *Distributional Hypothesis* ([Har68]) and the conclusion that semantic similarity is a function over global contexts ([MC91]). In other words similar words appear in similar contexts. In the presented experiments nouns are classified through their modifying adjectives or the verbs whose object or subject the noun is.

Chapter 2.1 outlines other works in the area of bootstrapping for semantic lexicons. Before the actual algorithm could be started several pre-processing steps such as semantically parsing the corpus, extracting subjects and their verbs or extracting significant co-occurrences had to be undertaken. They are explained in 2.2. Section 2.3 explains the algorithm and gives an example run. Afterwards the algorithm for combining the single features and ontological sorts is described in 2.4.

2.1. Other bootstrapping methods for the acquisition of semantic lexicons

Early bootstrapping works such as [RS97] were based merely on a couple of seed words and the simple context of one word left and one word right of the seed noun. This approach found new nouns in the given seed word category. The results were sorted by hand.

By separating [RS97]'s extraction patterns into two groups: conjunctions, lists, and appositives being one and noun compounds the other [RC98] gained significantly in precision since these two groups express quite different noun relationships.

Later more complex extraction patterns took into account a more sophisticated context. [RJ99] created the extraction patterns simultaneously with new nouns. Only with an

Chapter 2. Bootstrapping Algorithm

unannotated text and a couple of seed words it finds new words in categories such as *location, company* or *weapon* and uses the best 5 from each cycle to start another run. This mechanism is called meta-bootstrapping.

However high precision was difficult to obtain due to *result-infection*. Few words which are falsely classified in the beginning constantly deteriorated the results and thus the algorithm had to be stopped after a certain number of iterations.

[TR02] describes a bootstrapping algorithm that learns semantic lexicons for multiple categories such as *building, event* or *human*. It begins with an unannotated corpus and seed words for each semantic category. It then uses a large body of extraction pattern contexts to find multiple semantic categories simultaneously. An example extraction pattern that extracts people nouns would be

$$<subject> \; was \; arrested.$$

After each cycle only five new words are added to the lexicon and the process starts again, until no new words are found.

In a similar approach [AG00] tries to convert natural language information into relational tables by means of a meta-bootstrapping process called Snowball. The extraction rules are also evaluated after each step and only the best 5 are taken into the next iteration. An example pattern would be

$$<LOCATION>\text{-}based \; <ORGANIZATION>$$

which extracts the location of a company or organization together with its location and puts this pair into a database table.

In [PL05] parallel (as in word aligned) corpora together with shallow text are used for cross-lingual bootstrapping. The paper presents a fully automatic approach for semantic lexicon acquisition. It relies on the FrameNet lexicon and exploits word alignments to generate a list of frame candidates for a new language. The candidates are sorted out by a small set of linguistically motivated filters. Interesting aspects of this work are the fact that no deep syntactic or semantic analysis is needed and the comparison to the already mentioned SALSA (1.7) project. The result, a set of multilingual FrameNets for German and French, can be extended to other languages as well.

[BKW02] describes a method for the automatic classification of adjectives in German with respect to a range functional categories. They use the grammatical evidence that noun modifying adjectives only appear in coordinations if they belong to the same category and that there exists a relative ordering of adjectival modifier based on their functional category. Functional categories here are based on Engel [Eng88] and differentiate between quantitative, referential, qualificative, classifying and origin adjectives.

Nearest-neighbour similarity algorithms scale poorly with big vocabulary size and many different contexts as [Cur04] has shown. To reduce this problem a context-weighted approximation algorithm is introduced in [Cur04]. His idea is based on the fact that certain noun-verb relations are less indicative for a semantic similarity than others. One example for finding nouns of clothing are the verbs *get* and *wear*. The former appears in many different contexts whereas the latter is most often used with clothing nouns as their direct object.

This problem is considered and partly solved in this work by means of a parameter to the algorithm. Other aspects differentiate the approach in this work from the above mentioned. The lexicon in this work uses binary semantic features and ontological sorts which are bootstrapped separately in order to gain higher precision. Afterwards these are combined into complex semantic sorts. Relations are extracted from a semantically annotated corpus. Thus, the approach is not based on syntactical evidence or small context windows, but on semantic relations expressed with the MultiNet paradigm. In how far these changes result in high precision or not shall be discussed in section 3. Chapter 2.2 explains how polysemous nouns are treated, why predicate adjectives can be used as well and other details on the pre-processing steps.

2.2. Pre-processing Steps for Input Data

Scheme 2.1 shows the discussed pre-processing steps for a better overview. Basically the process starts with a flat corpus which is semantically parsed. Then a tool from the Fernuniversität Hagen is applied to extract certain relations. For each relation the following actions are taken:

1. create pairs in the form: *bootword - noun* (see section 2.2.2 for explanation)

2. delete personal pronouns

3. extract significant co-occurrences

4. a set with all co-occurrences (which can appear twice) is saved and one with only unique and significant co-occurrences, for each set the following measures are taken:

    a) one final input of the algorithm is a numbered bipartite graph with the co-occurrences, the numbers are just for easy access via a library

    b) an intersection with the noun database is performed, that means in the training set, only these nouns appear that are actually inside of the test or train set

    c) For each of the 34 semantic types (semantic sort, features and ontological sorts) train and test pairs are created for the bootstrapping process utilizes only one type at the time:

Chapter 2. Bootstrapping Algorithm

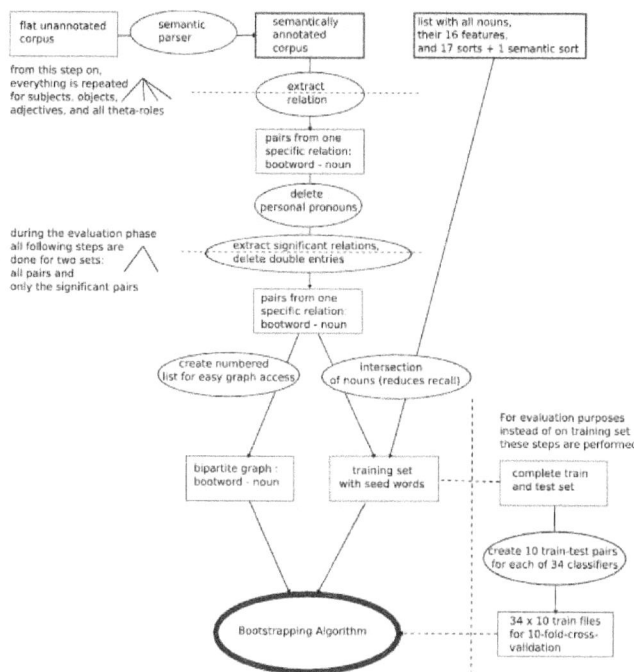

Figure 2.1.: Scheme with pre-processing steps for bootstrapping

  i. based on the 10-fold-cross-validation method 10 pairs for training and testing are created (see below for explanation)
  ii. each of these pairs together with the graph of the co-occurrences forms the input for the bootstrapping process.

During the actual usage of the described algorithm only one of the two sets (either all or only the significant pairs, based on their performance during the experiments) is used. Furthermore only one training set is created instead of 10 train-test-sets. These two differences need to be kept in mind when looking at figure 2.1.

2.2.1. Parsing the Corpus

Before describing the parser in this work, a short introduction to parsing in general is given, to understand the complexity of this task.

Chapter 2. Bootstrapping Algorithm

Parsing in NLP

The expression *parsing* is derived from 'partes orationis', that is part of speech. During the 19th century grammatical analysis circled mostly around finding the POS ([NL94]). This has changed dramatically since. Modern linguistics sees grammars as formal objects, which systematically represent phonological, morphological, syntactic, semantic and pragmatic information about natural language. Thus, a grammar has to fulfill the following criteria:

- formal adequacy, (i.e. consistency),
- empirical adequacy, such as adequacy of description, explanation or observation

A parser in NLP is often seen as a program that analyzes the structure of text input (phrases, sentences, texts) based on a formal grammar, tests whether the input is well formed with respects to this grammar and outputs an interpretation.

On the sentence level there are two important types, syntactic and semantic parsers. Sometimes these are used after a word-based morphological parser or before a text parser (that is, a system that analyzes discourse structure).

Often parsers are used for:

- testing for well-formed expressions
- find the structure for translation, question-answering systems or other processing
- disambiguate or identify the input
- correct the input
- improving and testing of a given grammar formalism

Unlike programming languages, natural languages have a very high complexity and structural diversity, many ambiguities, vagueness of terms and irregularities. As a result parsers in NLP are only able to analyze fragments of a given language, are very complex and have not yet achieved a full coverage of any natural language. The most important parsing methods shall only be mentioned here for completeness ([Car03]):

- shallow parsing, finds non-overlapping chunks of a phrase that are embedded in a flat structure
- dependency parsing, extracts head-modifier dependency links between words, which are labelled with the grammatical function of the modifying word (such as object or attribute)
- context free parsing, builds hierarchical phrase structure in the form of trees,

Chapter 2. Bootstrapping Algorithm

- unification-based parsing, uses attribute-value matrices to encode linguistic information and combines these with the unification operation based on hand coded rules, the most important example is head driven phrase structure grammar (HPSG), see [PS94]).
- parse disambiguation can be used after one of these tries to find the structure of a phrase. This is often based on tree banks. Tree banks are texts that are annotated by hand and which can be used to find the most probable version of several ambiguous readings of a phrase based on statistical evidence.

The WOCADI Parser

The WOrd ClAss based DIsambiguating (WOCADI) syntactic-semantic parser is a program from Fernuniversität Hagen written in Scheme ([Har03]). It translates German text into MultiNet representations and is being improved and employed in natural language interfaces, IR-systems on the internet ([JH02]) and question-answering systems ([Har05]).

WOCADI is one of the main applications that is based on the lexicon HaGenLex and is also used to verify its entries in example parses. **Thus, the extensions of HaGenLex which are based on this work help this parser to perform better.**

WOCADI first segments words and sentences of the input. Afterwards a morpho-lexical analysis is performed with the help of HaGenLex, another flat semantic lexicon and several proper nouns lexicons. This step returns the basic form for each word and its morphological information, such as declension or conjugation information. Since compound words are quite frequent in German a detailed analysis of these is also performed

The syntactic analysis is based on Word Class Functions ([HH97]). Results are represented in semantic nets of the MultiNet paradigm (see section 1.5.1).

During the presented experiments 42.1 % of all sentences of the corpus were parsed, the remaining sentences could not be used. The presented method of this work can be used in a cyclic iteration. A corpus can be parsed, bootstrapping on the results is performed and based on their outcome the parser's underlying lexicon is extended, which then improves its results and coverage.

Advantages of Semantic Parsing

The flat corpus is first parsed semantically. Doing this relatively expensive translation has several reasons. The first being that theta-roles can be extracted, also objects and subjects can be found even in passive constructions. It is one goal of this work to find out whether co-occurrences of verbs and nouns can be used to specify the noun semantically. Since the relation of an object (direct or indirect) and its verb is not very restrictive, results may not be satisfying. That is why the relation of theta roles is exploited, which

Chapter 2. Bootstrapping Algorithm

might yield better results. Furthermore predicate adjectives are treated the same as modifying adjectives. In addition collocations are found and treated as single units if appropriate. The verb 'stand out' for example can be found as one semantic unit.

However the most important reason is that polysemous words are contextually disambiguated. They are then annotated with numbers that represent each meaning and the bootstrapping process can work on the basis of these numbered nouns. See figure 2.3 for an example.

2.2.2. Extracting Context Relations

For the following experiments certain context relations are extracted from the semantically annotated corpus. Formally, such a context relation (or context) is a tuple

$$(b, r, n)$$

where

- b is the bootword occurring in
- a specific relation type r, with a noun n.

A bootword is an abstract term used for a certain type of context word that co-occurs with the nouns which are bootstrapped. During the experiments the following parts of speech are used :

- adjectives
- verbs.

Verbs hold different relations with nominal phrases in a sentence. This is taken into account by distinguishing between the subject and the object of a verb. Both of these are extracted from the corpus. This would also be possible by means of a syntactic analysis. To make an even more precise distinction of verb-noun relations theta-roles are extracted also. The relation r describes the particular relationship between the two words. The following context tuples are possible:

- (exciting, adjective, book)
- (walk, subject, dog)
- (letter, object, give)
- (write, theta-role:instrument, pen)[1]

---

[1] As mentioned before, the actual relations are in German. Notice also the solution to polysemy explained in the last paragraph.

Chapter 2. Bootstrapping Algorithm

In order to be able to calculate significant co-occurrences each instance (or occurrence) of a context is saved.

It is important to mention the possibility of extracting other relations as well. Simple nearest neighbour relations could be extracted even from a flat unannotated corpus and still be used as input for the algorithm. Though as previous experiments have shown those do not yield very high precision.

To illustrate how the extraction from the annotated corpus is performed an example query for the tool netarx from the Fernuniversität Hagen is shown. Netarx is the tool which extracts specific elements from semantic nets. Queries are written in a Prolog style language. The variable x_1 is a specialization of the verb of the sentence, a certain situation. The method of this situation is x_2. If this method is described by a noun, the verb and this noun are printed out:

```
'((subs ?x1 ?verb) (meth ?x1 ?x2) ((sub pred) ?x2 ?noun)
(cat ?noun n))' '?verb ?noun'
```

Figure 2.2.: A netarx query for the theta-role *method*

Fortunately there are no problems with polysemous nouns since the morphological parser also disambiguated between the different readings of a word and added a number for each meaning. One real example is shown in figure 2.3, notice that words are in their basic form and numbered with their corresponding meaning.

```
"kompromißlos.1.1"   "einhandkatamarane.1.1"
"akzeptabel.1.1"     "lebensdauer.1.1"
"bekannt.1.1"        "polit-punk-band.3.1"
"eigen.1.1"          "angabe.1.1"
"multipel.1.1"       "orgasmus.1.1"
"entscheidend.1.1"   "charakter.1.1"
"besonder.1.1"       "platz.1.1"
"öffentlich.1.1"     "telephonnetz.1.1"
"norwegisch.1.1"     "popband.3.1"
```

Figure 2.3.: Example with extracted adjective-noun pairs, showing numbers for each possible polysemous meaning

2.2.3. Extracting Statistical Significant Co-occurrences

For each specific type of relation the bootstrapping experiments are carried out based on two sets of bootword-noun pairs. The first is a list with all instances of context relations of the specified type, the other a list of only the significant co-occurrences from

Chapter 2. Bootstrapping Algorithm

the corpus. Unlike the former, the latter only contains unique pairs. Both are to be compared with regards to differences in precision and recall. The idea was that by using only significant co-occurrences (words that appear together for a reason) the precision would be higher, based on the fact that noisy combinations are deleted.

The term significant co-occurrence refers to a pair of two words, which occur together in specific positions. Their interdependence can be assumed with a certain minimum significance level. The significance level expresses the probability of a false rejection of the null hypothesis. In statistics, a result is significant if it is unlikely to have occurred by chance. In this case the result is the co-occurrence.

The first section explains why normal probability is unable to assure this level of significance correctly. The subsequent section explains how the log-likelihood method is used to filter out the significant co-occurrences. Lastly an explanation of the used threshold is given.

Co-occurrence Statistics

In a statistical test a hypothesis is postulated first. Then it is examined to which level of confidence this hypothesis can be accepted or rejected. The goal of co-occurrence statistics is to find out how surprising a given number of instances of a co-occurrence would be if the null hypothesis stated that the distributions were completely random (that is the words that form pairs were independent from each other).

Simple probability (p-values in [Moo04]) cannot capture the significance of rare co-occurrences correctly. Consider the following two examples:

1. Two words which both occur sixty times and co-occur half of the time together in a corpus with more than 10 million words. This is more surprising than if they had occurred only twice but together. Using the log-likelihood method this surprise can be discovered correctly even for small examples in megaword corpora. ([RC98])

2. Hapax legomenon words (singletons) are words that only occur once in the corpus. Since they are quite frequent in corpora, it is important to rely on measures which consider this fact. Given a huge number of hapax legomena, a certain number of them building pairs is normal due to chance. But these are usually only a small fraction (also called 'noise'), when compared to the actual number of hapax legomenon pairs. [Moo04] shows that seeing two hapax legomena together is really less significant than the use of standard probability values would suggest.

Both of these scenarios are based on sparse data in a big corpus. This is exactly the scenario where likelihood ratios are used as a significance measure instead of frequencies. Likelihood ratios are one approach for testing whether a hypothesis is true or not. The likelihood ratio can be seen as the confidence level for a specific hypothesis whose direction is calculated in a separate step. The bigger the likelihood ratio is, the more certain it is that a pair is significant and the two words are interdependent.

The Log-Likelihood Test

The log-likelihood score between verb-noun-pairs is computed according to [Dun93]. Following these outlined calculations the first step for finding the likelihood ratio of two neighbouring words A and B, is to calculate the following four values:

- n: Total number of bigrams (that is, a group of two words).
- n_A: Number of pairs in which A occurs.
- n_B: Number of pairs in which B occurs.
- n_{AB}: Number of pairs in which A and B occur together.

In step two the following equation is used to calculate the log-likelihood score for each pair[2]:

$$-2\log \lambda = 2 \begin{bmatrix} n\log n - n_A \log n_A - n_B \log n_B + n_{AB} \log n_{AB} \\ +(n - n_A - n_B + n_{AB})\log(n - n_A - n_B + n_{AB}) \\ +(n_A - n_{AB})\log(n_A - n_{AB}) \\ +(n_B - n_{AB})\log(n_B - n_{AB}) \\ -(n - n_A)\log(n - n_A) - (n - n_B)\log(n - n_B) \end{bmatrix}$$

λ denotes the likelihood ratio, that is the measure of surprise about the value of n_{AB} with a given n, n_A and n_B. A λ of 10 means, that one hypothesis is 10 times more probable that the other.

In order to find out if the number of co-occurrences n_{AB} is especially frequent or infrequent the probability of AB occurring together by chance (see null hypothesis in the beginning) needs to be calculated in step three with the formula:

$$P(AB) = P(A) \cdot P(B) = \frac{n_A}{n} \cdot \frac{n_B}{n}$$

If the number of actual co-occurrences n_{AB} is higher than the expected number based on this probability, then A and B are dependent and form a significant co-occurrence. With this calculation the direction of the confidence expressed in the likelihood ratio is clear and a statement about the null hypothesis is possible.

In step four a threshold c is introduced which is constant during all computations. If the value of step two ($2log\lambda$) is bigger then c and the null hypothesis is rejected based on step three, then the pair will be added to the set of significant co-occurrences.

---
[2]Based on personal communication with Stefan Bordag.

Chapter 2. Bootstrapping Algorithm

The Threshold of the Significance Level

Since the value of $-2\log\lambda$ is asymptotically distributed and similar to the χ^2-distribution, the χ^2 tables can be used for finding the significance ([MS99]).

The value of $-2\log\lambda = 6.63$ can be interpreted in the following way. With a probability of 99.0% the null hypothesis stating the independence of the two words can be rejected based on the actual instances. The probability of a wrong rejection or in other words, the level of significance is thus 1.0%. Based on the value of $-2\log\lambda$ the significance level can be calculated. The higher the significance (and thus also the computed likelihood-ratio), the more certain is the interdependence of two words.

For bigrams the table 2.1 can be used to find the significance level based on the $-2\log\lambda$ value. $P(correct\ rejection\ of\ H_0)$ refers to the probability of a correct rejection of the null hypothesis which states that the bigram occurred by chance.

Log-Likelihood Value	1.32	1.64	2.07	2.71	3.84	5.02	5.41	6.63	7.88	9.14	10.83
$P(correct\ rejection\ of\ H_0)$	0.75	0.80	0.85	0.9	0.95	0.975	0.98	0.99	0.995	0.9975	0.999
$P(false\ rejection\ of\ H_0)$	0.25	0.20	0.15	0.10	0.05	0.025	0.02	0.01	0.005	0.0025	0.001

Table 2.1.: Table for finding the significance level based on log-likelihood ratio

In the experiments of chapter 3 the threshold is at 3.84. With this value, a 5% error rate for finding significant *bootword - noun* pairs is defined. Furthermore two parameters introduced in section 2.3.2, try to minimize the effect of noisy co-occurrences.

2.2.4. Seed Words - Nouns in HaGenLex

Ontological Sorts for Nouns

As mentioned in section 1.5 there are 45 ontological sorts. But since the bootstrapping process is restricted to nouns during the experiments, only 17 of those are used. These are: object [o], concrete object [co], discrete object [d], substance [s], abstract object [ab], attribute [at], measurable attribute [oa], non-measurable attribute [na], relationship [re], ideal object [io], abstract temporal object [ta], modality [mo], situational object [abs], dynamic situational object [ad], static situational object [as], quantity [qn], unit of measurement [me].

For examples of each sort see table A.2 in the appendix.

Semantic Features

The following features are used for nouns or verb valency lists that restrict their possible inputs.

Chapter 2. Bootstrapping Algorithm

- *animal* - animal
- *animate* - living being
- *artif* - artifact
- *axial* - object having a distinguished axis
- *geogr* - geographical object
- *human* - human being
- *info* - (carrier of) information
- *instit* - institution
- *instru* - instrument
- *legper* - juridical or natural person
- *mental* - mental object or situation
- *method* - method
- *movable* - object being movable
- *potag* - potential agent
- *spatial* - object having spatial extension
- *thconc* - theoretical concept

For positive and negative examples of each feature see appendix table A.1.

Semantic Sorts for Nouns

Semantic sorts are the primary means to characterize words in HaGenLex. They combine each word with a predefined set of features and ontological sorts. Semantic sorts are treated differently since they are not binary. Furthermore do they subsume all features and ontological sorts in a hierarchy. In the beginning only two lists were available, one with a mapping of nouns to semantic sorts and another with semantic sorts and their subsequent characteristics. These were combined and one list was created for all 33 characteristics (the two sorts and the features).

A small excerpt of the 50 possible semantic sorts shall be presented in table 2.2 to get an idea of the principle.

Each of these is connected to set of characteristics. Exemplary shown for the semantic sort *human-object* in table 2.3:

This listing also shows that even though only HaGenLex is used during the experiments, the algorithm might as well work with any other characterization of the nouns, such as

Chapter 2. Bootstrapping Algorithm

abs−geogr, abs−info, abs−object, abs−potag, abs−situation, animal−object, animate−object, con−info, con−object, human−object, institution, meas−unit, modality, plant−object, prot−con−object.

Table 2.2.: Small excerpt of the 50 possible semantic sorts

human−object					
human:+	geogr:−	spatial:+	legper:+	instit:−	
animal:−	potag:+	movable:+	animate:+	info:−	
thconc:−	method:−	axial:+	mental:−	instru:−	
artif:−					
sort:d+	sort:na−	sort:abs−	sort:mo−	sort:ta−	
sort:co+	sort:ab−	sort:s−	sort:oa−	sort:io−	
sort:o+	sort:me−	sort:qn−	sort:ad−	sort:at−	
sort:re−	sort:as−				

Table 2.3.: Characteristics of the semantic sort *human-object*

freetime, *office* or any other category for nouns. For every single characteristic listed in table 2.3 a separate bootstrapping process is executed.

2.2.5. Bipartite Graph with Bootword-Noun Pairs

For easier access and fast computation co-occurrences are saved in the form of a bipartite graph. A bipartite graph is a special graph where the vertices can be decomposed into two disjoint sets such that no two graph vertices within the same set are connected with an edge. Formally this is:

A simple undirected graph $G := (V, E)$ is called bipartite if there exists a partition of the vertex set $V = V_1 \cup V_2$ so that both V_1 and V_2 are independent sets in the way that $V_1 \cap V_2 = \emptyset$ ([Die05]).

In this case the two sets are the bootword set and the noun set. Internally they are not connected. This setting is illustrated in figure 2.4

Chapter 2. Bootstrapping Algorithm

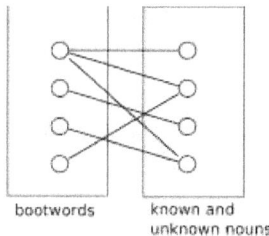

Figure 2.4.: Example of a bipartite graph

2.3. The Algorithm

2.3.1. Classification

The classification of new nouns is based on the profile of the bootword[3] that co-occurs with the noun. This alternation of profile calculation and new classifications is iterated in an Expectation-Maximization-Bootstrapping method ([DLR77]) until no new nouns can be classified. The algorithm is based on [BO05].

2.3.2. Principle

Overview

For each type of bootword and each semantic characteristic a separate bootstrapping process is started. After this initialization, the core loop alternates between two phases: During the first phase (training) a *class profile* is calculated for each bootword based on the number of its co-occurrences with nouns of a certain class. This profile expresses the probability of a bootword for each possible class.

During the second phase (classification) these *class profiles* are used together with *bootword profiles*. Each noun has a profile which saves all the bootwords this noun co-occurs with. The two different profiles are then employed to define which class each unknown noun shall inherit.

Based on the new classifications the *class profiles* are updated and the process starts again unless there were no new nouns classified in the last run. Figure 2.5 shows the algorithm in pseudo code.

---

[3] The artificial term 'bootword' refers to different possible parts of speech which co-occur with a noun: adjectives, and verbs. For the latter it is further differentiated between nouns that are objects, subject nouns and nouns in a specific theta-role.

Chapter 2. Bootstrapping Algorithm

```
Create necessary train file for one characteristic;
Initialize the training set;
Calculate bootword profiles;

While (new nouns get classified){
    Calculate class probabilities;
    For (each unclassified noun n){
        Multiply class probabilities class-wise;
        Assign class with highest probability to noun n;
    }
}
```

Figure 2.5.: Bootstrapping algorithm for assigning semantic attributes to nouns

Initialization Phase

During the initialization, bootword profiles are listed for all nouns in the bipartite graph. These profiles do not change during the process. Inside the graph known and unknown nouns are mixed. Bootword profiles have the form

$$noun_x -> bootword_1, bootword_2, \ldots, bootword_n$$

Furthermore seed words in the training set are connected with their class:

$$noun_y -> class_1$$

Learning Phase

The following steps are iterated in the outer loop as long as new nouns get classified. Each bootword gets assigned class probabilities that express how indicative this word is for all classes. The probability is calculated for each class from its frequency distribution, divided by the total number of nouns in that class (relative frequency). All class probabilities for this bootword are then normalized to one.

Division by the total number of nouns is motivated by skewed class distributions where one class has only a fraction of the size of another. Since bootwords gain new class information due to newly added nouns, class probability needs to be recalculated after each step.

Classification Phase

Having these values new classes are assigned to hitherto unclassified nouns by multiplying class probabilities of all bootwords from the bootword profile. These same class probabilities are multiplied only from bootword that had occurred inside a profile of a known noun.

The new noun inherits the class with the highest probability. Note that if one of the values is zero, that is the corresponding bootword never occurs with this class, a class cannot be assigned. Smoothing would give each class a minimum pre-defined probability. This measure has shown smaller precision without a considerable gain in recall though and is thus not used as default.

After the algorithm ends, many new nouns can be classified based on the results.

Parameters

There are four important parameters settings that can alter the outcome of the algorithm:

1. **maxClass** - maximum number of classes per bootword: This parameter is especially important when it comes to complex characteristics that can have more than two different values. As [Cur04] showed there are verbs like *get* that do not hold very much information about their objects. Similar examples for adjectives (such as *good*) exist too. With the help of this parameter, these can be filtered out. In case of a binary characteristics there are only two meaningful values: 1, for a very strict version that does not allow inheritance of classes with bootwords that co-occur with nouns of different classes, or 2, which is the normal case where good results can still be obtained as the experiments in section 3.2.2 show.

2. **minBoot** - minimum bootwords per noun: A new class gets only assigned if there are at least this many bootwords in the profile of a noun.

3. pre-processing the input: Filter for significant co-occurrences or compute with all pairs

4. smoothing: Each class gets a minimum default probability, so one bootword in a profile cannot result in a 0 chance for the noun to inherit this bootword's class.

Standard values for the first two are (2 5), but numerous other possibilities are tested during the experiments (3). Especially the pre-processing steps and **minBoot** are analyzed, since they changed the results the most during the adjective - noun tests.

2.3.3. Example

This subsection shall demonstrate a small example with real German pairs to better understand the algorithm. While the experiments are only performed for German, the algorithm can also be utilized for other languages.

Chapter 2. Bootstrapping Algorithm

The algorithm begins with a set of pairs, figure 2.6 shows an excerpt from verb-object relations. The first set of pairs is formed with nouns whose classes are known, the other with unknown nouns. Thus, the nouns *mensch* (human) and *maler* (painter) need to be added to the lexicon.

werden.1.2	weltverbesserer.1.1	human:+
werden.1.2	individualist.1.1	human:+
werden.1.2	atheist.1.1	human:+
lieben.1.1	weltverbesserer.1.1	human:+
lieben.1.1	individualist.1.1	human:+
lieben.1.1	jazz.1.1	human:−
werden.1.2	mensch.1.1	?
werden.1.2	maler.1.1	?
lieben.1.1	maler.1.1	?
überraschen.1.1	maler.1.1	?

Figure 2.6.: An excerpt from German verb-object relations

Another more abstract visualization of these pairs is shown in figure 2.7:

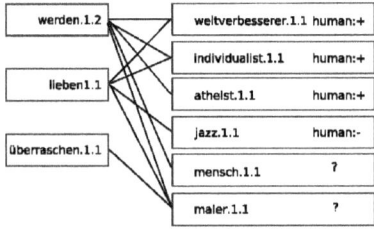

Figure 2.7.: Abstract visualization of an excerpt from German verb-object relations

First the distribution is calculated:

werden.1.2:	3+	0−
lieben.1.1:	2+	1−

Then the distribution is divided by the total number of nouns in that class.

werden.1.2:	3/3+	0/3−
lieben.1.1:	2/3+	1/3−

Both relative frequencies are normalized to one, by calculating for *lieben* (love):
$p(+) = \frac{\frac{2}{3}}{1+\frac{2}{3}} = \frac{2}{5}$ and $p(-) = \frac{1}{1+\frac{2}{3}} = \frac{3}{5}$

30

Chapter 2. Bootstrapping Algorithm

werden.1.2:	1.0+	0.0−
lieben.1.1:	0.4+	0.6−

Now the nouns get classified with these probabilities:
For *mensch.1.1* the probabilities are: $p(+) = 1$ and $p(-) = 0$, so it is sure that *mensch.1.1* is *human:+*.

For *maler.1.1* it is:
$p(+) = 1 \cdot 0.4 = 0.4$ and
$p(-) = 0 \cdot 0.6 = 0$
So *maler.1.1* also gets the class *human:+*. Notice here that the verb *überraschen.1.1* could not be used since it has not yet appeared in any bootword profile of a known noun.

This section described the core bootstrapping method in all its details including the numerous possible pre-processing steps and the parameters. An example demonstrated the theory of the algorithm for a faster understanding of the concept and the parameters during the testing.

2.4. Assignment of Complex Semantic Sorts

After the bootstrapping finished for the 33 binary characteristics, these can be combined which showed significantly better results than bootstrapping directly on the 50 complex semantic sorts ([BO05]). The process of their combination is described in this section. The actual program was provided by Rainer Osswald from the Fernuniversität Hagen.

Figure 2.8 shows the method for combining characteristics (features and ontological sorts) to semantic sorts. The first operation inside the for-loop assigns possible values for characteristics which have not been bootstrapped for the specific noun.

```
for each new noun{
    list all semantic sorts, which are compatible with single characteristics;
    extract the sorts that are minimal in the specialization relation;
    if (one element is left){
        this element is the resulting sort;
    }
    else {
        no sort is given to this noun;
    }
}
```

Figure 2.8.: Algorithm for combining characteristics to complex semantic sorts

Thus, the characteristic is weak since the algorithn does not assign a result in each case. Section 3.2.2 shows experimental results of this combination method.

Chapter 2. Bootstrapping Algorithm

2.4.1. Summary

This chapter explains the bootstrapping algorithm with possible pre-processing steps that prepare the input data, the core algorithm and its parameters and the combination of the results.

The pre-processing steps described are relatively expensive, which is why it is important to mention that other methods for retrieving possible input data do exist. Syntactical analysis for example is possible for finding subject nouns, or in the case of adjectives nearly no analysis is needed, if only the adjectives directly occurring before a noun are used.

However the morphologic-semantic analysis does have advantages such as the extraction of prepositional adjectives, the direct contextual disambiguation of polysemous nouns and verbs and the extraction of theta rules. Furthermore the extraction tools allow to only include parses above a certain threshold of certainty of correctness.

The following experiments in chapter 3 are based on the lexical semantic structure of HaGenLex which provides a greater differentiation of noun description. However the algorithm can again be used to bootstrap any other set of lexical features without any adaption in code.

Chapter 3.

Experiments, Results and Interpretation

This chapter covers all experiments on the algorithm. Numerous runs with different parameters are performed to gain the best results. Each type of relation is separately analyzed and compared with other types.

The extraction from Wikipedia and the parsing was performed at the Fernuniversität Hagen by Rainer Osswald and Sven Hartrumpf. The core bootstrapping algorithm is written in Java and the tool to calculate significant co-occurrences in Perl, both by Christian Biemann at the Universität Leipzig. The bootstrapping program has been modified to allow different input, such as the verbal relations. For the extraction of relation pairs, the tool netarx from Hagen was used. To automate the numerous extractions, pre-processing steps and to create numbered list for the bipartite graphs, perl scripts were programmed by the writer of this work, Richard Socher. For fast access in co-occurrences in graphs a library by Stefan Bordag at the Universität Leipzig was incorporated ([BBH+04]). To automate the bootstrapping process for all relations bash scripts were produced in the course of this thesis. The automatic analysis of all relations with all different parameter settings was programmed in Perl and the R-language (mostly for graphs) by Richard Socher. The combination to semantic sorts is done by a program from Rainer Osswald.

The first two sections cover the basics of the analysis, the corpus, extracted pairs and used measures. The first run of experiments discussed in section three uses algorithm settings and parameters from past experiments ([BO05]) for easier comparison of the new relations. The remaining sections use these settings as a starting point but also analyze possible changes. Furthermore more complex analyses are carried out that go beyond precision and recall values. These are explained in section 3.2.

Chapter 3. Experiments, Results and Interpretation

3.1. Corpus, Extracted Co-occurrences and Noun Database

3.1.1. Results from Wikipedia and CLEF Corpus

The starting point for the experiments of this work are two different corpora:

1. 230,223 different texts from the German version of the community encyclopedia Wikipedia[1] were extracted in May 2005, resulting in a total of 4,482,526 sentences.

2. The German Cross Language Evaluation Forum (CLEF) Corpus with 276,579 different texts and 4,809,171 different sentences. It consists of newspaper articles from 1994 and 1995.

As described in section 2.2, these corpora are parsed semantically into MultiNet's semantic representations. 42.1% of all sentences were parsed, the remaining sentences could not be used. To improve the coverage by enlarging the number of known nouns is one of the goals of this work.

After parsing, pairs of bootword-noun co-occurrences are extracted. Table 3.1 shows statistics of extracted pairs from each relation. A short definition of the terms is needed before, when talking about adjective-relations what is actually meant is the relation between the adjective as the bootword and the modified noun. On the other hand *subject* refers to the relation between a verb and this verb's subject. The same holds for objects and the theta-roles. The first column shows the number of all pairs, that means the same pair might occur more than once. To contrast that and to prevent an overestimation of the significance test, the second column shows how many of these pairs are unique in the corpus. Based on column one the significant co-occurrences are calculated and presented in column three. The last two columns show how many unique nouns and bootwords were combined to pairs. The first three rows show the major relations, the rest shows the relations of verbs to nouns in specific theta-roles. For an explanation of the theta-role abbreviations, see section 1.4.

The following relevant correlations have been found. First of all even though there are about half a million more total adjective-noun pairs than verb-noun(object) pairs, the number of unique pairs from the latter is still higher. More unique relation instances then result in more significant co-occurrences following the hapax legomenon pair theory outline in section 2.2.3. Seemingly contrary to that, there are less unique words (both in object and subject relations). Thus, the high number of significant pairs in subject and object relations is not based on many unique single words, but rather on the fact that more words occur in different pairs. Also interesting to note is that there are more unique adjectives in the corpus than there are verbs, which is common in most corpora. The possible problem with theta-roles is also evident. In most verb-noun (in a certain

---

[1] see http://www.Wikipedia.de

Chapter 3. Experiments, Results and Interpretation

Type of Relation	Number of Pairs	Number of Unique Pairs	Number of Significant Pairs	Number of Different Nouns	Number of Different Bootwords
Adjectives	3,769,485	778,217	675,613	192,939	62,377
Subject	3,372,498	817,157	698,379	209,821	15,554
Object	3,278,012	797,454	690,072	220,407	10,668
θ-obj	1,933,688	487,776	421,065	160,990	23,503
θ-agt	1,883,870	404,691	324,583	102,086	18,427
θ-benf	577,910	161,062	140,444	48,667	14,633
θ-exp	413,235	111,324	98,226	48,177	8,496
θ-ornt	374,243	100,067	86,881	43,631	4,680
θ-aff	350,625	91,425	83,359	43,089	5,728
θ-mcont	299,761	81,150	69,088	34,994	4,532
θ-rslt	215,445	52,472	47,710	33,031	1,868
θ-cstr	145,787	51,249	46,997	23,456	7,563
θ-oppos	82,618	23,850	21,505	11,870	2,570
(θ-instr)	58,099	21,637	20,797	9,021	5,161
θ-avrt	36,081	9,626	8,643	6397	500
(θ-init)	20,195	6,969	6,282	5136	261
(θ-supp)	16,167	804	422	734	45
(θ-meth)	4,249	1,770	1,739	1157	786

Table 3.1.: Pre-processing: number of extracted pairs from Wikipedia and CLEF corpus

theta-role) relations there are not many pairs in the corpus. Results of experiments are shown in section 3.2.2.

Since more than three million pairs are available only for adjective-, subject- and object-relations these are talked about separately and referred to as the 'main groups'. Theta-roles are subsumed after examining whether they work in a bootstrapping approach even with fewer co-occurrences.

3.1.2. Possible Drawbacks of the Wikipedia Corpus

There are three possible reasons why this corpus might not be optimal for this approach. First the language is not always sophisticated. Often simple words and simple sentence structures are used. Another problem is that articles often create 'small worlds'. This means that some words only occur inside one article and thus do not co-occur often with other words from other articles. Last but not least the corpus could be larger. Solving these problems with a bigger and better corpus, such as the whole *Projekt Deutscher Wortschatz* (PDW) corpus (see next section) is not possible due to resource and time constraints. The general applicability of this approach can however still be demonstrated and better results are highly probable as shown with a small subset of the PDW corpus.

Chapter 3. Experiments, Results and Interpretation

3.1.3. Corpus: Projekt Deutscher Wortschatz

For testing a hypothesis made by experiments based on these two corpora, a subset of the corpus *Project Deutscher Wortschatz*[2] (PDW) was analyzed also. This corpus was created under the supervision of Uwe Quasthoff in 1994. There are over 36 million German sentences (as well as 10 million English and Catalan sentences and smaller corpora for other languages) from newspapers, encyclopedias and other texts ([QRB06]). Stress was placed on a balanced corpus from different sources and fields. Each word entry has the following information (see [QHW98], [QBW02]):

- absolut frequency,
- synonyms and antonyms,
- grammar information,
- semantic category,
- example phrases and
- significant co-occurrences

9 million sentences were extracted from the sentence database and parsed semantically with WOCADI. In total 695316 verb-noun(subject), 638,194 verb-noun(object) and 773,333 adjective-noun pairs were extracted.

As mentioned in section 2.2 about the pre-processing steps, the other main input is a list of nouns and their classes. All experiments started with a set of 11,100 nouns, their features, ontological and semantic sorts. Though not all of these nouns occurred in the extracted relations or in significant pairs, thus, the recall is also restricted to what is left after an intersection of both noun sets.

3.2. Experiment Settings

3.2.1. Experiment Measures

During the experiments the 10-fold-cross-validation method is used. This method partitions the initial set of known seed nouns into ten subsamples, retaining class distribution. For all testing purposes, nine subsamples are used for training and the remaining for testing. Then the next subsample is used for testing and the other nine are for training, this is done until all subsamples were used for testing exactly one time. So in the end the whole set of available nouns was used for verification and can be used to calculate precision and recall.

---

[2]see http://www.wortschatz.uni-leipzig.de

Chapter 3. Experiments, Results and Interpretation

Precision is the ratio of found relevant (correct) tokens over the total number of found tokens (correct and wrong).

$$precision = \frac{number\ of\ relevant\ found\ tokens}{total\ number\ of\ found\ tokens}$$

The ratio of found relevant tokens (correct ones) over the total number of tokens in the search domain.

$$recall = \frac{number\ of\ relevant\ found\ tokens}{total\ number\ of\ tokens}$$

To combine these two values, the weighted harmonic mean of both called F_α-score is used, for non-negative real α it has the general form of:

$$F_\alpha = \frac{(1+\alpha) \times \text{precision} \times \text{recall}}{\alpha \times \text{precision} + \text{recall}}$$

For the actual extension of the high quality semantic lexicon, the precision of the results is more important than the recall, thus, $\alpha = 0.5$ has been chosen, which weights precision twice as much as recall.

$$F_{0.5} = \frac{1.5 \times \text{precision} \times \text{recall}}{0.5 \times \text{precision} + \text{recall}}$$

In past experiments only the set of known nouns was analyzed. But also newly classified unknown nouns can bear useful information beyond their sheer amount. One measure introduced in this work that is based on unknown nouns is called *doubt*. During each of the ten training-bootstrapping-testing runs unknown nouns are classified. Since only one tenth of the training set differs, most of the nouns get classified more than once. However, with a slightly different training set, different classification occurs, if the characteristic is unstable. This is reflected in the *doubt* measure, which is calculated by dividing the smaller number of a found binary class of a noun by the bigger number. Doubt is zero, if the same class was given to the noun in all ten[3] runs. This measure works only on binary characteristics. A short definition is:

Doubt is a ratio for binary characteristics that reflects the doubt of classifications by comparing the results from different training sets. To find general facts for wrong classifications, average doubt is listed as the arithmetic mean over all single word doubts. Specific word examples that gave wrong results are also given in section 3.7.4.

3.2.2. Parameters for First Experiments with the Three Major Groups

The motivation of the distinctions between these two groups is influenced by the *Distributional Hypothesis* ([Har68]) and the conclusion that semantic similarity is a function

---

[3] If the n-fold-cross-validation-method is used with another n, then this is reflected here, too.

Chapter 3. Experiments, Results and Interpretation

over global contexts ([MC91]). With this in mind, one might ask *In how far can a verb be used to semantically identify the nouns in its valency list?*. Answering this question is one of the goals of this section and its experiments.

The identified four main settings of past experiments with adjectives ([BO05]) are used as a starting point for the presented experiments in this work.

1. pre-processing the input: only significant co-occurrences form the input
2. smoothing: no smoothing on class probabilities
3. maxClass: maximum number of different classes for each bootword is set to 2
4. minBoot: minimum number of bootwords in bootword-profiles of unknown nouns is set to 5

For the results to be comparable with regards to the relation, these parameters are not changed during the first set of experiments. The goal is to find out whether adjective modifiers or certain verb - noun relations are more useful for clustering nouns. After this fundamental correlation has been identified, all parameters and the input are changed, differences are analyzed and subsequent experiments performed accordingly. Since changes in smoothing have not shown to result in big changes, these experiments are only briefly compared.

For the three major groups (called adjective, for adjective-noun-pairs, object, for verb-noun(object)-pairs and subject, for verb-noun(subject)-pairs), the two bootstrapped characteristics (features and sorts) are thoroughly analyzed and results of their combination to complex semantic sorts is given. In the end combining graphs and scores are shown for better comparison. So for the three main relations the following subsections are used to organize the analysis:

1. analysis of features with tables of all single results, means and two graphics
2. analysis of ontological sorts with tables of all single results, means and two graphics
3. analysis of both of these sets of characteristics to semantic sorts

If the tables are not in the corresponding section, then they can be found in the appendix.

Theta-roles play a separate role since fewer pairs could be extracted from the corpus. Theta-roles are thus, described in groups that show similar behavior.

The recall in this and all other tables about binary characteristics is always based on the 11,100 nouns in the noun database. For the three major groups the maximal possible recall is around 80 % because only about 80% of the 11,100 known nouns actually occur in the pairs on which the bootstrapping is performed. Bias values are however calculated based on the number of actual occurrences of each class in the pairs, since these numbers are more representative, especially for very skewed classes.

Chapter 3. Experiments, Results and Interpretation

3.3. Adjective - Noun

Adjective-noun co-occurrences and their bootstrapping results were explained in detail in [BO05], this section compares those results. Differences might result from the varying approaches in extracting the pairs or the corpora themselves.

Two simplified extraction examples shall demonstrate what kind of relations are used in this section.

1. *This story is hilarious.* ⇒ hilarious - story (predicative)
2. *I like this interesting thesis.* ⇒ interesting - thesis (adjunct)

All words in this and other parts are in German and contextually disambiguated.

3.3.1. Features

First the feature distribution of noun classes in adjective-noun pairs is listed in table 3.2. Column four shows the bias, which is used for binary characteristics to show the ratio of the bigger class. It is calculated by counting the number of occurrences of both classes and dividing the bigger number by the sum of both. The last three columns display the results of the bootstrapping algorithm. Precision is the most important in this setting, though recall should not be too low either. It shows how much new information can be gained and thus, in how far the lexicon can be extended, based on the input given. The f-score combines both measures, with an emphasis on precision. For further information on these measures consult section 3.2.

To get a picture of the consequences of a bias that is too high, table 3.2 also shows precision for the smaller positive class (labelled with 'Precision:+'). Because the feature *spatial* occurs more often positively than negatively the smaller negative class is used for calculation. The overall mean of smaller positive classes is at 73.4%. The mean in the table does not consider the size of the classes, but instead uses the precision values of all features evenly. Overall the precision for positive classes is high enough to use the results, though some features are not suitable for this method. For *method* for example precision is only at 1.0, because the recall is 0, thus no new nouns were falsely assigned the positive class.

The problem with bias above a threshold of about 95% exists in all other experiments with features and sorts and thus it is looked for well balanced characteristics.

With the exception of *artif* the precision is above 82%. Notice also that the maximum f-score value is at 0.77. The overall precision and recall is dependent on the bias, as seen in figure 3.1. The y-axis shows precision and recall values that depend on bias which is always shown on a scale from 0.5 to 1.0 on the x-axis. The regression line (also called least-squares line) shows an estimation of the expected value of one variable y

Chapter 3. Experiments, Results and Interpretation

Characteristic	Number+	Number-	Bias △	Precision	Precision:+	Recall	F-score
spatial-	4629	4333	0.5165	0.8609	0.8393	0.3625	0.5904
axial	3694	5266	0.5877	0.8510	0.8517	0.3633	0.5880
movable	3632	5330	0.5947	0.8432	0.8581	0.3605	0.5830
potag	2935	6024	0.6724	0.9003	0.8798	0.3930	0.6294
artif	2791	6170	0.6885	0.7590	0.5744	0.3262	0.5263
animate	2698	6256	0.6987	0.9120	0.8954	0.4063	0.6446
legper	2102	6852	0.7652	0.9154	0.8293	0.4103	0.6490
human	2039	6915	0.7723	0.9205	0.8416	0.4154	0.6550
instru	1285	7671	0.8565	0.9053	0.6252	0.4386	0.6683
thconc	705	8248	0.9213	0.8242	0.1626	0.4085	0.6154
animal	408	8540	0.9544	0.9786	0.7948	0.5234	0.7587
geogr	262	8686	0.9707	0.9406	0.1549	0.4825	0.7145
mental	197	8751	0.9780	0.9622	0.1610	0.5176	0.7480
info	173	8775	0.9807	0.9316	0.0031	0.4871	0.7143
instit	63	8885	0.9930	0.9872	0.0243	0.5287	0.7659
method	24	8924	0.9973	0.9955	1.0000	0.5391	0.7764
Mean	-	-	0.8092	0.9055	0.5310	0.4352	0.6642

Table 3.2.: Adjectives: features, distribution, bias and bootstrapping results

(here precision and recall) given the values of another variable x (here bias), mostly this estimation is given in the form of $y = \alpha + \beta x$.

It is preferred to have a high precision even for smaller bias values and thus a less steep regression line.

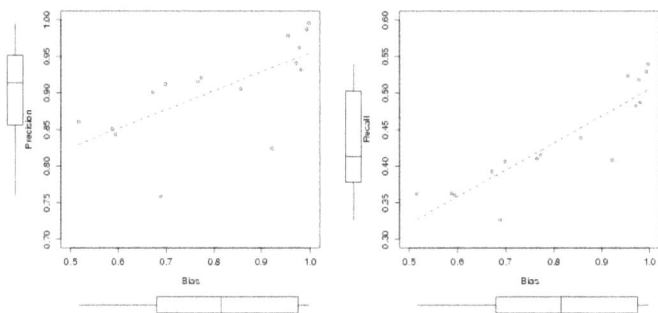

Figure 3.1.: Adjective-noun: features - precision and recall

The recall is highly dependent on the bias and only at the maximum of 55%, if the bias is almost 1. The small box-whisker-plot left of the precision graph shows the precision meridian at 91.4%. The overall precision mean is at 90.6%. This is 3.3% worse than in [BO05]. Recall is at 43.5 %, half of the maximal possible recall. This value cannot be compared to the average recall of 70 % in [BO05] though, since the corpus in this work is significantly smaller. Other problems with the corpus are outlined in section 3.1.2.

For the characteristics *animal, geogr, mental, info* and *method* nouns with a positive class are rarely occurring and the bias is too high. In these cases the great precision results only from allocating the negative class almost exclusively. The feature *instit* for example classifies 41 of 63 possible positive class occurrences, but only one of them is identified correctly as an institution, all other 40 were thought to be negative. The same holds for *geogr*, with a precision of only 15% for positive values and method with 0%.

3.3.2. Ontological Sorts

Ontological sorts show similar connections in table 3.3 and figure 3.2. Due to a higher bias, precision (mean: 92.8%) and recall (mean: 47.1%) are slightly higher. For the precision of positive classes, the two sorts *co* and *o* are used inversely, since the negative class forms the bigger set. In order to show how a bias that is too high, deteriorates the results for the smaller class, the precision for the smaller class is also listed in the column 'precision:+'. For the sorts sort:o-, sort:mo, sort:oa, sort:me, sort:qn, sort:re, sort:na, sort:at, less than 10 elements were found at all, but these were false. These sorts are rarely represented in the corpus and a manual verification is not problematic, since very few nouns got classified. The overall precision for smaller classes which takes into account the sizes of each individual class is at 70.0 %. The low value of 0.3250 comes from using small classes equally and only calculating the mean over all single means. In the cases of *o* and *co*, the negative class is used for calculations since it is rarer.

Characteristic	Number+	Number-	Bias △	Precision	Precision:+	Recall	F-score
sort:co	4629	4333	0.5165	0.8609	0.8392	0.3625	0.5904
sort:d	4289	4672	0.5214	0.8511	0.8643	0.3584	0.5836
sort:ab	4247	4709	0.5258	0.8668	0.8419	0.3659	0.5953
sort:abs	2752	6199	0.6925	0.8281	0.7037	0.3685	0.5849
sort:ad	2266	6684	0.7468	0.8189	0.6251	0.3739	0.5863
sort:io	928	8025	0.8963	0.7705	0.1913	0.3657	0.5628
sort:as	486	8463	0.9457	0.8778	0.1553	0.4525	0.6684
sort:at	417	8531	0.9534	0.9754	0.4166	0.5253	0.7587
sort:na	379	8569	0.9576	0.9792	0.25	0.5306	0.7639
sort:s	288	8661	0.9678	0.9732	0.3835	0.5135	0.7495
sort:ta	127	8821	0.9858	0.9862	0.2549	0.5274	0.7645
sort:o-	8900	70	0.9922	0.9960	0	0.5428	0.7792
sort:qn	48	8900	0.9946	0.9960	0	0.5430	0.7793
sort:me	48	8900	0.9946	0.9960	0	0.5430	0.7793
sort:oa	38	8910	0.9958	0.9950	0	0.5405	0.7772
sort:re	11	8937	0.9988	0.9985	0	0.5438	0.7809
sort:mo	11	8937	0.9988	0.9979	0	0.5441	0.7808
Mean	-	-	0.8638	0.9275	0.3250	0.4707	0.6991

Table 3.3.: Adjectives: sorts, distribution, bias and bootstrapping results

Notice that in both graphs, the regression line is quite steep, showing a high dependence of both values on the bias.

3.3.3. Combination to Semantic Sorts

The binary characteristics are combined with the method described in section 2.4. In the case of adjectives all semantic sorts are possible, since adjectives can describe situations as well as concrete nouns, methods, etc.

The overall precision of semantic sorts is at 62.0%. The two biggest groups in the corpus (with about 2500 occurrences) are also the ones with the highest precision: *nonment-dyn-abs-situation* and *human-object*. The former is a non-mental, dynamic, abstract

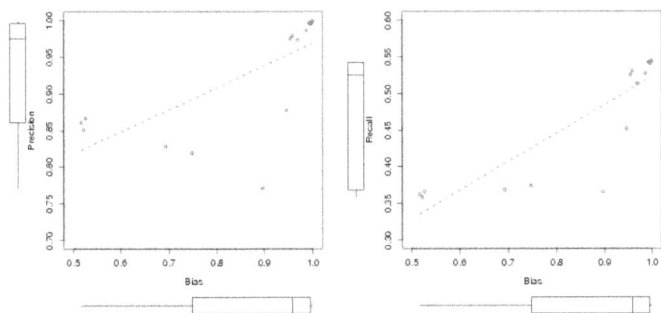

Figure 3.2.: Adjective-noun: sorts - precision and recall

situation, such as the concept of 'marriage'. Only a few other sorts have a precision above 60%, such as *animal-object* or *tem-abstractum*. For a list of all sorts that occur more that 100 times in the corpus (or make up more than 1% of the nouns) see appendix table B.1, these are only 20 of 50 possible classes, the remaining occur very rarely or not at all. The precision is neither dependent on the distribution of the semantic sorts nor on the total number of occurrences.

Since the final outcome of this process is very important for extending the lexicon, the overall precision for adjectives is not satisfying. Compared to the experiments in [BO05], it is also below expectations of about 80%. Besides the same restrictive method of combination, this is mostly due to fewer extracted pairs i.e. a smaller corpus and the problems related to the Wikipedia corpus in general.

3.4. Verb - Noun(Subject)

The term *subject* is a vague concept. The subject of a sentence can be a noun, a phrase, a gerund, a clause etc. Ergative-absolutive languages such as the Basque language do not declare a subject. For German this term is justified syntactically and semantically. The *subject* is the word or phrase which holds the focus and forms a complete sentence together with a verb. Normally it is situated at a certain position in the flat or syntactically parsed tree structure of a sentence. In MultiNet representations it is always the first complement of the verb. For bootstrapping only single nouns in this position are extracted together with their verb.

A simplified extraction example shall demonstrate what kind of relations are used in this section.

Chapter 3. Experiments, Results and Interpretation

1. *The monkey, who sits in front of a typewriter, ate that interesting thesis for lunch.*
   ⇒ eat - monkey

The interesting question is: *Can a verb be used to find information about its subject?*. Of course there are some verbs that cannot, such as *is*. But as this section shows, verbal relations, which have hitherto not been been analyzed in this environment show significant improvements in precision.

3.4.1. Features

Even though these pairs have a bias mean of around 80% like adjective-noun relations, they show a 4% increase in precision combined with a 1% higher recall. Table 3.4 shows that again a problem with the *artif* class exists, which is going to be analyzed in section 3.12. One very interesting aspect is that even with almost balanced classes such as *spatial* or *axial* the precision is still above 90%.

Characteristic	Number+	Number-	Bias △	Precision	Precision:+	Recall	F-score
spatial-	4791	4022	0.5436	0.9628	0.9707	0.4358	0.6862
axial	3882	4929	0.5594	0.9041	0.9236	0.4052	0.6410
movable	3774	5039	0.5718	0.8905	0.9041	0.3986	0.6309
potag	3142	5668	0.6434	0.9844	0.9747	0.4370	0.6945
animate	2898	5905	0.6708	0.9727	0.9547	0.4308	0.6854
artif	2581	6231	0.7071	0.8370	0.6787	0.3868	0.6030
legper	2211	6593	0.7489	0.9631	0.8896	0.4258	0.6779
human	2148	6655	0.7560	0.9588	0.8818	0.4233	0.6744
instru	1271	7536	0.8557	0.9133	0.6467	0.4366	0.6696
thconc	666	8135	0.8870	0.9243	0.3306	0.4446	0.6661
animal	464	8333	0.9473	0.9732	0.7654	0.4866	0.7299
geogr	258	8539	0.9707	0.9382	0.1644	0.4704	0.7046
mental	190	8607	0.9784	0.9512	0.1355	0.4833	0.7192
info	170	8627	0.9807	0.9226	0.0771	0.4640	0.6939
instit	63	8735	0.9928	0.9900	0.1363	0.4986	0.7452
method	21	8777	0.9976	0.9912	0	0.5064	0.7514
Mean	-	-	0.8030	0.9400	0.5896	0.4459	0.6858

Table 3.4.: Subjects: features, distribution, bias and bootstrapping results

The resulting regression curve is very flat as seen in figure 3.3. This is very desired since it shows a high robustness of the process and demonstrates that the approach works well with balanced classes. The problem with small positive classes is obvious in table 3.4, because precision for the smaller positive class is below 20 % if the bias is above 95%.

With the same bias mean, verb-subject-relations show a significantly better result in precision than adjective-noun-relations.

3.4.2. Ontological Sorts

The same higher precision can be seen in the experiments with sorts. Even for balanced class distributions the score is above 90 % in four cases. The fact that the overall precision is even better than for features is a result of many classes having a skewed distribution as table 3.5 shows. The sort:re is only 1 because no class was assigned at all.

Chapter 3. Experiments, Results and Interpretation

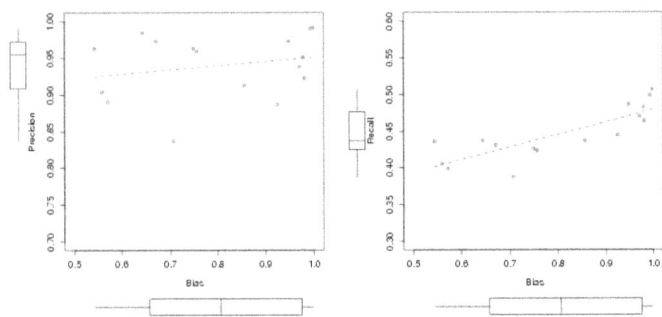

Figure 3.3.: Verb - noun(subject): features - precision and recall

Characteristic	Number+	Number-	Bias △	Precision	Precision:+	Recall	F-score
sort:d	4446	4366	0.5045	0.9343	0.9457	0.4187	0.6624
sort:co	4791	4022	0.5436	0.9628	0.9707	0.4358	0.6862
sort:ab	3940	4864	0.5525	0.9620	0.9543	0.4384	0.6881
sort:abs	2607	6192	0.7037	0.9218	0.8156	0.4367	0.6727
sort:ad	2142	6657	0.7566	0.8858	0.7161	0.4351	0.6585
sort:io	893	7909	0.8985	0.8683	0.3995	0.4236	0.6432
sort:as	465	8332	0.9471	0.8523	0.1583	0.4308	0.6427
sort:at	318	8479	0.9639	0.9825	0.3157	0.5009	0.7440
sort:s	295	8503	0.9665	0.9638	0.3350	0.4824	0.7233
sort:na	285	8512	0.9676	0.9882	0.3333	0.5045	0.7489
sort:ta	111	8686	0.9874	0.9771	0.0389	0.4959	0.7383
sort:o	8755	65	0.9926	0.9988	0.9987	0.5106	0.7574
sort:me	42	8755	0.9952	0.9988	1	0.5106	0.7574
sort:qn	42	8755	0.9952	0.9988	1	0.5106	0.7574
sort:oa	33	8764	0.9962	0.9917	0	0.5059	0.7512
sort:mo	11	8786	0.9987	0.9981	0	0.5102	0.7568
sort:re	0	8797	1.0000	1.0000	1	0.5113	0.7584
Mean	-	-	0.8688	0.9579	0.5283	0.4742	0.7145

Table 3.5.: Subjects: sorts, distribution, bias and bootstrapping results

Even though the overall precision seems to be so great only due to the high bias, this does not hold true since even balanced sorts such as *d, co, ab* and *abs* all show above 92% precision.

The extremely high bias is also seen in the meridian of the marginal box plots beneath the scatterplots shown in figure 3.4. This is a sign of bad precision for the smaller class. Indeed, for the *at sort* which stands for all kind of non-measurable and measurable noun-attributes such as *charm* or *length* the +-precision is at 31%. Only 38 were given a new class, out of which 12 were correctly assigned the *sort:at+*. Other sorts with an above 94% bias show similar +-precision.

3.4.3. Combination to Semantic Sorts

The same method that was used for adjectives is applied to the bootstrapped characteristics in this section. As a direct result of a higher precision of both features and

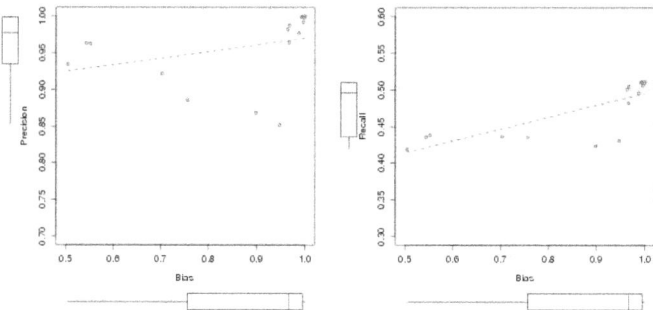

Figure 3.4.: Verb - noun(subject): sorts - precision and recall

ontological sorts the results of their combination show improvements too. With an increase of 13% to an overall precision of 75.8% it is significantly higher than that of the adjective results. As table B.2 in the appendix shows most of the semantic sorts have a higher precision than with adjectives. Especially good results are again for the two biggest groups of *nonment-dyn-abs-situation* and *human-object*. Furthermore other typical subjects such as nouns referring to the concept of an *animal-object* or of a theoretical concept give good results. Considering the loss in precision seen in the adjective results, a bigger corpus could improve the results for subjects, which tremendously increase the quality of the output.

3.5. Verb - Noun(Object)

The term *object* is in the same manner vague as the term *subject*. There is a distinction between direct and indirect object. For the purpose of these experiments, only the second verb complement in the semantic nets of MultiNet are extracted. Whether the object is a direct or indirect object thus depends on the verb.

Simplified extraction examples shall demonstrate what kind of relations are used in this section. For a more thorough understanding, the actual German sentences are listed also.

1. *If a bird cannot swim it is not a penguin.* ⇒ is - penguin
   Wenn ein Vogel nicht schwimmen kann, ist er kein Pinguin. ⇒ sein - Vogel

2. *The children accuse the teacher of writing boring exams.* ⇒ accuse - teacher
   Die Kinder werfen dem Lehrer vor, langweilige Klausuren zu entwerfen. ⇒ vorwerfen - Lehrer

Chapter 3. Experiments, Results and Interpretation

3. Many books are read by some children ⇒ read - book
   Viele Bücher lesen manche der Kinder. ⇒ lesen - buch

3.5.1. Features and Ontological Sorts

Since the results are very similar to those of the subject bootstrapping, only the average values are presented here. For complete tables see appendix tables B.3 for features and B.4 for sorts. The overall mean values are listed in table 3.6.

Interestingly, the most balanced features such as *spatial*, *axial* and *movable* show very high precision (between 90% and 98%) and so do well balanced sorts such as *co*, *d* and *ab*. With only a 1% increase in bias, the precision mean and the recall for features are 5% higher than in the adjective experiments. While the precision increases only 1% compared to verb-subject-relations, recall gained another 4%.

The same performance gain can be seen for sorts.

Characteristics	Bias	Precision	Precision:+	Recall	F-score
Features Mean	0.8179	0.9508	0.5950	0.4877	0.7217
Sorts Mean	0.8642	0.9642	0.6312	0.5101	0.7430

Table 3.6.: Objects, mean values: features and sorts, distribution, bias and bootstrapping results

The problem with very small classes in some characteristics persists. See table B.5 in the appendix for a detailed list with the actual number of found correct words and precision. But these low values represent the precision of the positive class which in some cases occurs less than 50 times in 8000 words (for example, *re, mo, qn, me, oa*). These values are significantly higher than the corresponding adjective values (0.5310 for features and 0.3250 for sorts).

Figure 3.5 shows that for all bias values of the different features, precision is above 90% (except, *thconc* with 89.4%). The regression line shows almost no dependance on bias, which is desired for a robust bootstrapping method.

The two graphs in figure 3.6 demonstrate similar results, with a minimum precision is of 86.2%.

3.5.2. Combination to Semantic Sorts

For object bootstrapping only 3 sorts have a worse precision as seen in table B.6. The remaining 14 sorts show better results. Table 3.7 shows the seven best results, together with their fraction of all nouns, the number of times they were correctly and falsely identified, their recall and the precision. It is observable that distribution does not

Chapter 3. Experiments, Results and Interpretation

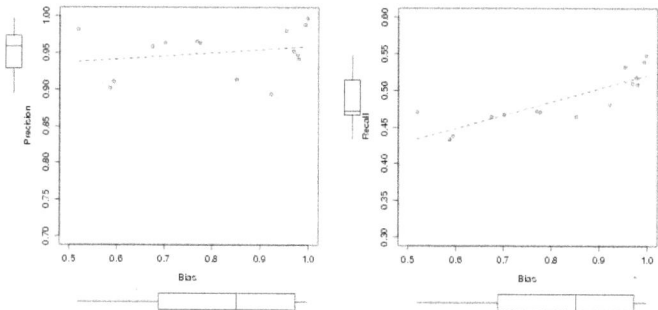

Figure 3.5.: Verb - noun(object): features - precision and recall

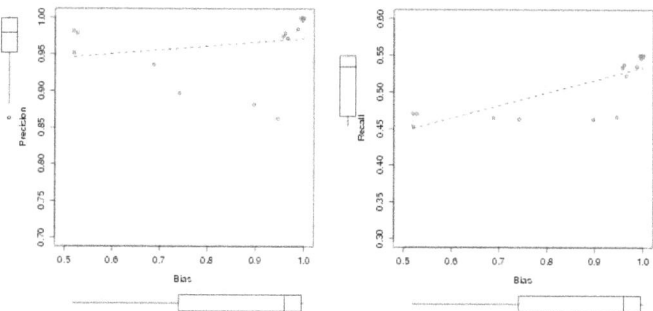

Figure 3.6.: Verb - noun(object): sorts - precision and recall

influence precision value for this method.[4] The average precision over all semantic sorts is 73.4%.

Some semantic sorts do not occur in any of the three main groups:

- nonthc-relation
- attribute
- abs situation

Which suggests that these are not very important.

---

[4]Though, indirectly, since the method for combining is based on the results of the classification of binary features and ontological sorts.

Semantic Sort	Total	%Distribution	Correct + Wrong	Correct	%Recall	%Precision
human-object	2359	21.26	1079	997	42.26	92.40
animal-object	593	5.34	36	33	5.56	91.67
nonment-dyn-abs-situation	2751	24.79	961	815	29.63	84.81
abs-info	148	1.33	37	27	18.24	72.97
tem-abstractum	143	1.29	29	21	14.69	72.41
prot-theor-concept	815	7.34	128	82	10.06	64.06
mov-nonanimate-con-potag	159	1.43	28	17	10.69	60.71

Table 3.7.: Objects: Combination to semantic sorts - best results

3.6. Verb - Noun(Theta-Role)

The notion of theta-roles and their theory has been introduced in section 1.4 and 1.5.5. They provide a more fine grained way to classify nouns in a sentence and elements in valency lists of verbs. The reason why bootstrapping might work better with only nouns in a specific theta-role instead of nouns subsumed under the vague notions of *subject* or *object* is that the nouns are already sorted semantically before the bootstrapping process starts. This effectively reduces the number of times where verbs co-occur with totally different nouns.

Take for example the theta-role *benf*, which is a relation between a situation or an object and another object. Beneficients are always concrete objects, if not they would have a different theta-role. By bootstrapping only on certain semantic elements that are still diverse internally, some characteristics might be inherited more correctly. This section wants to find out if this is the case. In addition more pairs can be extracted since theta-roles can be in different positions of verb valency lists (and not only the first complement as is the case for extracted object).

MultiNet defines 15 different theta roles. Some of these are very rare in corpus, other occur very frequently. To find out which theta-roles might be the most interesting to compare, all relations were bootstrapped and analyzed, with results as seen in figure 3.7. This is already the first part of the comparison, though necessary to understand the choices for more detailed analysis. The graph shows the average precision, bias and f-score of all features and sorts that were bootstrapped ordered by relations and their f-score. As mentioned in section 1.5.5, the roles *meth, instr, init* and *suppl* are only listed for completeness and not analyzed.

Special attention was put on theta-roles that result in balanced characteristics. These have almost the same amount of positive and negative elements of a class and can thus be used to find both groups correctly.

Chapter 3. Experiments, Results and Interpretation

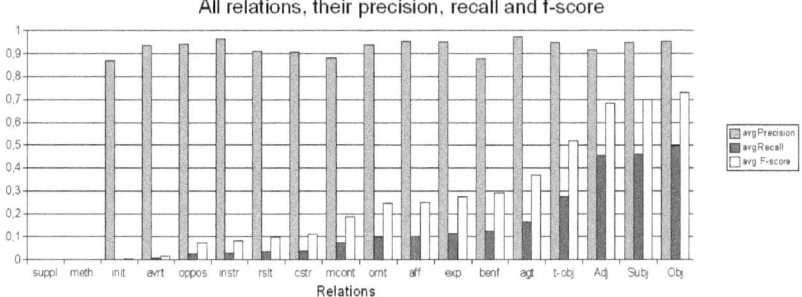

Figure 3.7.: F-score of all relations

3.6.1. Features and Ontological Sorts

This section describes the behavior of all theta-roles with regards to their single characteristics.

Though not obvious in graph 3.7 one can see the strong dependence of high f-scores from the number of extracted pairs in table 3.8.

Relation	Number of Pairs	Average F-score	Average Precision	Average F-Score
object	3,278,012	0.7296	0.9542	0.4969
subject	3,372,498	0.7006	0.9492	0.4605
adjective	3,769,485	0.6822	0.9168	0.4535
θ-obj	1,933,688	0.5192	0.9474	0.2733
θ-agt	1,883,870	0.3692	0.9745	0.1647
θ-benf	577,910	0.2908	0.8777	0.1247
θ-exp	413,235	0.2742	0.9500	0.1134
θ-aff	350,625	0.2484	0.9548	0.1002
θ-ornt	374,243	0.2461	0.9381	0.0994
θ-mcont	299,761	0.1853	0.8830	0.0718
θ-cstr	145,787	0.1100	0.9069	0.0399
θ-rslt	215,445	0.0966	0.9114	0.0347
θ-oppos	82,618	0.0730	0.9404	0.0257
θ-avrt	36,081	0.0172	0.9339	0.0058

Table 3.8.: Correlation between number of extracted pairs and average F-score

The three major groups are not ordered by the number of pairs, which suggests that object-relations work best in this environment. However the f-score correlates to the number of pairs used for bootstrapping. This demonstrates the strong dependence on big corpora. It is important to notice that this dependence on the number of pairs is mostly due to the steadily declining recall (actual and maximal possible) and not precision, which is only three times below 90%.

Theta-Roles: mcont, benf

Since the goal is to find semantic groups for effective bootstrapping, only those with high precision are analyzed in more detail. For the effective theta-roles to find more new

Chapter 3. Experiments, Results and Interpretation

nouns, is then only a matter of a bigger corpus. The lowest precision with 88% shows θ-mcont, the few high precision values for some sorts are only based on unbalanced distributions in this case. Even though many pairs were extracted for θ-benf, the precision is one of the worst. Not a single class with a balanced distribution shows good precision values, which suggests that the theta-role of *beneficent* is not suitable for this method. These are the three cases with a precision below 90%.

Theta-Roles: cstr, rslt, avrt, ornt, oppos, obj

The following evaluation looks in all remaining theta-roles for characteristics that have an above 90% precision with bias below 85% (preferably around 50-70%). Since a small bias is an indicator for high precision for both classes (positive:+ and negative:-) and high precision with above 90% bias is only the result of assigning the bigger class, as explained for the three main groups. However some classes are so rare or even nonexistent in a certain theta-role that this information can be used as well.

With an average precision of 90.7% the theta-role for '**causator**' (*cstr*) is the least precise in the remaining set of theta-roles. High precision is mostly caused by skewed class distributions, some of which even allow general conclusion. If a noun is found to be in the theta-role of *cstr*, it can be stated that it

- does not have the positive feature *mental*
- is not of sort *re* (relationship), *as* (static situational object, such as sleep), nor of the sorts *qn* and *me* (quantity and measurement).

The theta-role **result** (*rslt*) shows a balanced distribution, combined with a high precision for the features *animate, legper, human* and is thus a good candidate to bootstrap these. Similarly table 3.9 lists certain sorts and features that show very high precision with a balanced distribution for other theta-roles. The theta-roles are ordered by average precision, beginning with the lowest.

Theta-Role	Features and Sorts with very High Precision
θ-rslt	animate, legper, human
θ-avrt	sort:io, sort:ab, instru
θ-ornt	potag, animate, human, instru, thconc
θ-oppos	sort:co
θ-obj	animate, legper, sort:d human, sort:co, spatial, instru

Table 3.9.: Theta-roles and some single characteristics with above 90% precision

Theta-Roles: aff, exp, agt

The remaining four theta-roles have an average precision of above 95% for all characteristics. They can be partitioned into two classes:

Chapter 3. Experiments, Results and Interpretation

1. *aff* (affected objects) and *exp* (experiencer) both have some well distributed characteristics which show very high precision and most of the unbalanced characteristics show even for the positive classes high quality results.

2. *agt* (agent) shows very high precision and high recall, since it bootstraps on many pairs.

The theta-role of affected objects can be used to bootstrap features and sorts listed in table 3.10 especially well (*exp* specifies a subset of these characteristics in a similar fashion).

Features/Sorts	Bias	Precision
potag	0.5687	0.9311
animate	0.6106	0.9514
legper	0.6878	0.9500
human	0.6997	0.9512
spatial	0.7734	0.9767
sort:d	0.7001	0.9073
sort:co	0.7734	0.9767
sort:ab	0.7796	0.9787
sort:io	0.8394	0.9351

Table 3.10.: Theta-Role *aff*: High precision values

Probably the best theta-role suited for bootstrapping in this approach is *agt*. Many pairs could be extracted from the corpus, so the recall is comparably high, the semantic pre-classification is beneficial for this method (method, mental, sort:re and sort:as never occur positively), very small classes get either classified correctly or not at all and balanced distributions also result in above 90% precision. For detailed results see table B.7 in the appendix.

3.6.2. Semantic Sorts

Since the *agent* theta-role is relatively easy to extract in a parser and one of the few theta-roles that shows high precision and recall for all characteristics it is examined in more detail. This analysis is quite short, since the theta-role of *agent* occurs almost exclusively with one semantic sort: *human-object*. Of 1481 correctly specified semantic sorts 1424 are in it. These are found with a 90.0% precision. The remaining sorts are mostly *animal-objects* (41 correctly identified with a precision of 97%) and *mov-nonanimate-con-potag* (14 found correctly, precision 93%). For finding these semantic sorts it is fruitful to bootstrap on the theta-role *agent*.

The role of affected objects (*aff*) works well for combining *mov-nonanimate-con-potag*, *tem-abstractum*, *animal-object*, *human-object* and *animate-object*. All with a precision of above 70%.

Chapter 3. Experiments, Results and Interpretation

3.7. Comparison of Adjective-Noun Relations and Verbal Contexts

This section compares the bootstrapping results for binary features of the four main groups (with a note on theta-roles) and thus gives an answer to the general question whether verbal or modifying (i.e. adjectival) contexts are more useful for this bootstrapping method. The combination to semantic sorts is based directly on these results, thus the conclusion can be extended to their domain.

3.7.1. F-score of the Three Main Groups

To easily compare the three main relations, features and sorts are combined in one graph and recall and precision are merged into the f-score (see 3.2). Since all three results are based on almost the same amount of pairs ($\approx$ 3.3 million) and has a similar average bias the comparison does not need to take these into account. However the fact that this amount resulted in these findings in the first place needs to be kept in mind when looking at theta-roles. The combination of these measures can be seen in figure 3.8. On the left side adjective bootstrapping results can be seen. The regression curve is very steep, indicating a high dependence on skewed class distributions. The smallest f-score is at 0.53. Compared to the flat regression lines of the subjects graph in the middle and the objects graph on the right, it can be stated that adjective results are more dependent on bias and less robust. Furthermore verbal relations show a higher precision as seen

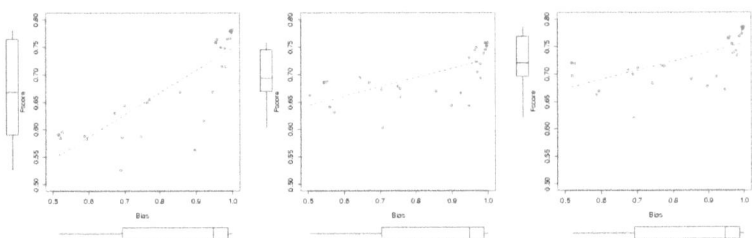

Figure 3.8.: Adjective, subject and object f-scores, based on bias

in the corresponding sections 3.4 and 3.5. Adjective-experiments have a precision of 90.5% for features and 92.7% for sorts, subject-experiments have 94.0% and 95.8% and object-experiments show the best outcome with 95.1% (features) and 96.4 % (sorts). Not only is the overall precision higher for objects, but each single feature and sort gets better values too.

Chapter 3. Experiments, Results and Interpretation

3.7.2. Verification and Conclusion

Since a recent study ([HOT06]) found out that modification works better for finding synonyms than subject or object relations, this is a surprising discovery. Several external reasons might account for this difference. Since it is a corpus based approach a closer look at the corpus is necessary. One thing that comes to attention is that while there are more unique nouns in verb-noun(object) relations than in adjective-noun relations (220,000 vs. 192,000), there are many more corresponding unique adjectives (62,000) than there are unique verbs (11,000). In addition there are slightly more unique verbs in verb-subject relations (16,000). Thus, fewer verbs occur with more nouns. The following hypotheses are stated to account for these correlations:

- More adjective-noun pairs might form disconnected subgraphs in the bipartite graph on which the algorithm performs the bootstrapping. Nouns which are part of such a pair cannot inherit new classes.

- Since the number of significant pairs is in an equal range of about 680,000 it can be deduced that each adjective occurs less frequently in general and with a certain class. Thus, the certainty with which they speak for one class decreases[5]. As a consequence their performance as a medium to transfer classes from one noun to another decreases.

- Lastly the chance that one adjective in a profile never occurs with a class and thus rules out all possibility of passing this class to its noun (even though all other elements speak for it) is higher.

However, these differences alone cannot account for the higher precision for objects and subjects as experiments with the *Projekt Deutscher Wortschatz* (PDW) show. For the three main groups, the bootstrapping was repeated. Even though the ratio of unique words: $\frac{|bootwords|}{|nouns|}$ is different in these experiments[6], the results are still in favor of verbal relations as table 3.11 demonstrates. It shows the number of unique bootwords and nouns, their ratio and the received precision.

Relation	Number of bootwords	Number of nouns	Ratio	Precision
PDW: Adjective	5913	16458	0.359	0.927
PDW: Object	4291	11595	0.370	0.968
PDW: Subject	4798	9851	0.487	0.970

Table 3.11.: Projekt Deutscher Wortschatz: Bootstrapping precision for adjective, object and subject relations

---

[5] If the edges are relatively balanced between train and test set.
[6] The experiments with the different corpus are preformed under exactly the same circumstances and with the same parameters.

The subject relation, having the biggest ratio, gets the highest precision. In the analyzed experiments before, adjectives with a ratio of 0.323, had worse precision than objects (ratio: 0.048) and subjects (ration: 0.07).

In conclusion verbal relations show higher precision and better overall results in this bootstrapping approach no matter how high the ratio in the bipartite graph is.

3.7.3. Combination to Semantic Sorts

The results of the combination to semantic sorts are quite low in all experiments. Since parameters are the same in this section as in ([BO05]) and in both cases adjectives are used, the differences are mainly due to the corpus or the parser. In order to prove that the new results are independent of the corpus[7], **semantic sorts were also combined on the basis of the subset of the PDW corpus**. Experiments with a higher precision for binary features resulted in a more precise combination. Adjectives receive the lowest semantic sort precision with 67%, while **subjects can be used more effectively (86%) than in all other experiments**.

This outcome expresses two important conclusions, because it receives better results with a smaller number of pairs:

1. A well balanced corpus is very important for the combination to semantic sorts.

2. Verbal relations show better results independently of the corpus.

Some semantic sorts in HaGenLex define many features and sorts very specifically. If for example a semantic sort is based on one characteristic being positive, but this positive characteristic occurs hardly ever ($< 1\%$) in subjects or the first complement of verbs, then this semantic sort is underspecified in the combination process. No classification can be performed for underspecified nouns right now. Some semantic sorts are not found in relatively basic language used in Wikipedia and can thus not be bootstrapped at all. These problem do not exist for binary features, which provide important information themselves.

3.7.4. Doubt-Measure

The *doubt*-measure (see 3.2) indicates, that of all unknown nouns, the most robust new classifications were performed by the subject relation. The sum of the doubt-means of all characteristics is only at 0.16 in comparison to 0.22 for objects and 0.50 for adjective. This way *doubt* can express the relative robustness of several relations. In order to improve results, only words with a zero doubt could be used.

---

[7]Independent in the sense that verbal relations are more effective than adjectival relations, not in the sense that both results are independent from the corpus.

Chapter 3. Experiments, Results and Interpretation

By extracting a small set of example words with the maximum doubt value of 1 and the minimal doubt of 0, its usefulness shall be outlined. However complete tests with overlapping test sets would prove more useful because they enable a correlation between doubt and actual precision. This goes beyond the scope of this work but should be investigated in the future.

Figure 3.9 shows extracted, arbitrarily chosen, newly identified example words and their doubt values and the bootstrapped value of the class. For the *animate* feature, words which have been assigned each class 5 times are often related to human beings such as a police car (Streifenwagen) with humans inside, or a Mediterranean island (Mittelmeerinsel) where humans live or spend their holidays. The flowering plant (Blütenpflanze) is certainly animate. The *artificial* feature listing shows theoretical concepts created by humans, such as a statute (Artikelgesetz) or abstract nouns, such as 'household' with a high doubt. But the term 'temporary constitution' (Übergangsverfassung) which is also in that group has a zero doubt. It is interesting to note that 'town hall' (Stadthalle) and public finances (Staatsfinanz) have high doubt values twice. In summary, further studies are needed to employ this measure effectively.

```
animate:
1:plaque.1.1| 1:blütenpflanze.1.1| 1:mittelmeerinsel.1.1| 1:handicap.1.1| 1:streifenwagen.1.1|
1:impression.1.1| 1:wikisource.1.1|
0:verteilschlüssel.1.1(-)| 0:sparkonzept.1.1(-)| 0:übergangsverfassung.1.1(-)| 0:stadthalle.
1.1(-)| 0:staatsfinanz.1.1(-)| 0:gemeindewappen.1.1(-)|

artif:
1:studentenverbindung.1.1(-)| 1:sportvereinigung.1.1(-)| 1:haushalt.1.1(-)| 1:ölgeschäft.1.1(-)|
1:verordnungsentwurf.1.1(-)| 1:artikelgesetz.1.1(-)| 1:entwicklungsphase.1.1(-)| 1:bankgeschäft.1.1(-)|
0:sparkonzept.1.1(+)|0:übergangsverfassung.1.1(+)| 0:stadthalle.1.1(+)| 0:staatsfinanz.1.1(-)|
0:gemeindewappen.1.1(+)| 0:pauschalbetrag.1.1(+)| 0:jugendkeller.1.1(+)|
```

Figure 3.9.: Example doubt values

3.7.5. Theta-Roles

Most theta-role experiments have the problem of too few extracted pairs. While *theta-object* is well suited in general, it is not that different from the general notion of an *object*. The most successful theta-role, which is also easy to extract in a sentence is *agent* (*agt*). It is the only relation that successfully bootstrapped the feature artifact (*artif*, 95% precision), it has an above 90% precision for all characteristics and a recall of more than 50% of its maximal possible (17% of 29%). However it is only suitable to find four semantic sorts, since others sorts do not occur in that theta-role. Generally theta-roles are suitable to find certain single characteristics very well (see section 3.6), but due to low recall on some characteristics and their semantic restrictions these cannot be combined to semantic sorts successfully.

One possibility that goes beyond the scope of this work, but should be examined in the future is to combine single characteristics from each relation while ignoring others.

3.8. Filtering for all Significant Co-occurrences

The three sections 3.8, 3.9 and 3.10 demonstrate different results based on parameter changes for the three main groups. Theta-roles behave the same way. In the beginning all experiments were performed with two sets:

1. the set of all pairs, including multiple entries and
2. the set of significant pairs without multiple entries,

to find out whether this rather expensive step is really improving the results. It does improve the outcome, but only marginally. Some characteristics constantly have higher precision based on the first set. Figure 3.10 shows the results of bootstrapping with the three major relations. Each graph shows two groups, the green triangles represent precision values obtained from significant pairs. Notice how these are above the red circles representing the results of the normal set. In case the latter indicates better results, it is marked with a number for explanation afterwards. These are for example the *sort:ad* (59, dynamic-situational-objects), the feature *thconc* (21, theoretical concept) or *sort:as* (65, static-situational-objects). This is most likely due to the deletion of pairs that would have inherited the classes correctly, but were deleted, since they were not significant.

It is interesting to note that for the set of all relations, edges in the bipartite graph are weighted, since they can occur more than once for equal pairs. For example if the pair 'woman - see' occurs 30 times and we know that woman has the feature 'human:+', then 'see' gets this feature 30 times in his bootword profile. For this to happen with significant pairs, 'see' would need 30 *different* 'human:+' nouns in his profile. Because overall results of binary characteristics are slightly worse for these weighted graphs, it can be stated that deleting double entries and bootstrapping on unweighted co-occurrences shows higher quality solutions. One reason is that very large classes tend to make other smaller classes impossible to inherit.

In summary it can be stated, that filtering for significant co-occurrences is necessary to obtain the highest results. However, in environments were time and resources are important factors, it can be left out, since it improves the overall precision only about 1%.

3.9. Effects of Smoothing

Smoothing was identified as one of the main parameters. Without it, one bootword in a profile, which never occurs with a single class, is enough to prevent this class from being assigned. Starting with the default parameters and switching smoothing on with a minimum probability for each class of 0.001, did not result in major changes. Average

Chapter 3. Experiments, Results and Interpretation

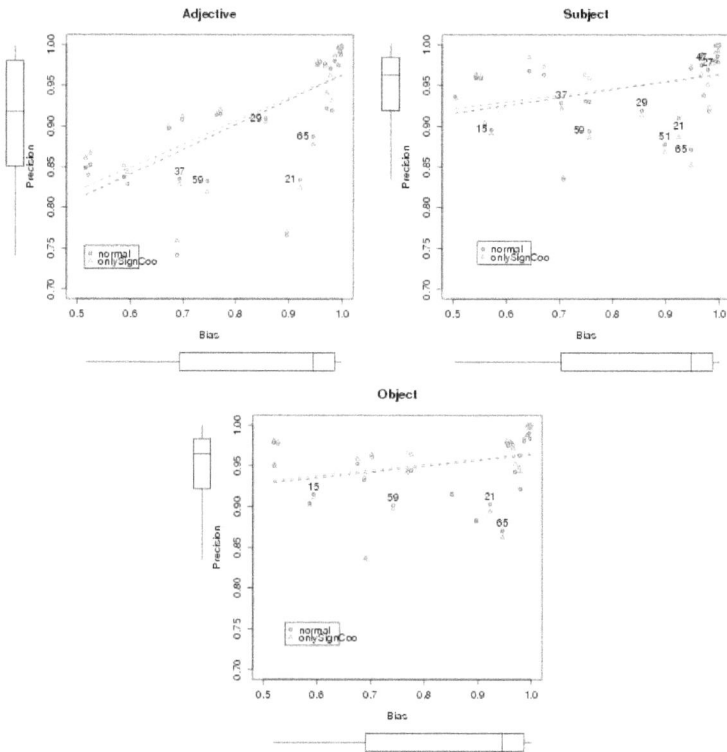

Figure 3.10.: Different results based on filtering for significant pairs.

precision only decreased less than 1% for most relations and recall gained from 0% to 5% for some.

To sum up, one can say that with a 1% decrease in precision and 1% increase for recall (in the mean of all relations, including theta-roles) smoothing is not effective.

3.10. Effects of Parameter Changes: minBoot and maxClass

As mentioned in section 2.3.2 the last two parameters are maxClass (the maximal amount of classes, a bootword can have, before it is forbidden to pass a class) and minBoot, the

Chapter 3. Experiments, Results and Interpretation

minimum number of elements in a bootword profile. Of course, for binary characteristics used in this method, there are only two meaningful values for maxClass:

1. only bootwords that occur exclusively with one class are used to pass on new classes

2. a bootword can occur with both classes, and passes the class with the higher probability

Two main outcomes can be formulated. Firstly with $maxClass = 1$ precision improves significantly, but recall drops. Secondly, the same effects can certainly be stated for $minBoot$ values above the default of 5, though it will turn out less strongly. In past experiments the $maxClass$ parameter was neglected, but its strict limit bears the possibility of very high precision as table 3.12 shows. This table compares results of the 3 main groups, based on changes to the parameters $minBoot$ and $maxClass$. Smoothing is disabled and only significant co-occurrences are used. Since objects return the best results they are listed on the last line of each parameter group. This table bears many

maxClass	minBoot	Relation	Precision	Recall	F-score
1	1	adjective	0.9351	0.5151	0.7148
1	1	subject	0.9756	0.5651	0.7759
1	1	object	0.9765	0.5939	0.7944
1	2	adjective	0.9655	0.3888	0.6020
1	2	subject	0.9886	0.4341	0.6756
1	**2**	**object**	**0.9894**	**0.4636**	**0.6987**
1	3	adjective	0.9742	0.3173	0.5221
1	3	subject	0.9928	0.3585	0.6018
1	**3**	**object**	**0.9930**	**0.3875**	**0.6281**
2	2	adjective	0.8948	0.5873	0.7608
2	2	subject	0.9239	0.6082	0.7873
2	**2**	**object**	**0.9324**	**0.6391**	**0.8083**
2	3	adjective	0.9045	0.5369	0.7352
2	3	subject	0.9357	0.5499	0.7580
2	3	object	0.9437	0.5828	0.7818
2	4	adjective	0.9120	0.4918	0.7083
2	4	subject	0.9435	0.5027	0.7296
2	4	object	0.9500	0.5370	0.7557
2	5	adjective	0.9168	0.4535	0.6822
2	5	subject	0.9492	0.4605	0.7006
2	5	object	0.9542	0.4969	0.7296
2	6	adjective	0.9207	0.4176	0.6550
2	6	subject	0.9538	0.4247	0.6733
2	6	object	0.9573	0.4643	0.7064
2	10	adjective	0.9293	0.3136	0.5589
2	10	subject	0.9645	0.3118	0.5672
2	10	object	0.9643	0.3602	0.6177

Table 3.12.: Results with different parameter values of maxClass and minBoot

interesting information. First of all, the highest recall is at 2-2 (pairs in that form are always of type $maxClass - minBoot$), which is not surprising because two co-occurring bootwords are enough and both can even occur with different classes. In environments where high recall is more important and a precision of around 92% is sufficient this value pair is certainly the best choice.

For extending a high quality semantic lexicon, precision is of highest importance, since all data that enters the lexicon needs to be verified. **Thus, in summary the pair 1-3 applied to object-relations provides the best solutions for binary features.** With an average precision of 99.3% for binary characteristics, it provides results that can be put into the lexicon with very little supervision.

Chapter 3. Experiments, Results and Interpretation

In addition, this combination returns the highest precision after combining the binary features to semantic sorts. The following 10 semantic sorts show an above 50% precision (in comparison, there are only 6 for adjectives) in the object-experiments and the pair 1-3: *nat-discrete, nonment-stat-abs-situation, plant-object, abs-info, prot-theor-concept, nonment-dyn-abs-situation, tem-abstractum, human-object, animal-object, nat-substance.*

Furthermore the *doubt* is always zero for results where the parameter maxClass was set to 1, which suggests higher quality solutions for all new unknown words.

However for the combination to complex semantic classes, recall values need to be high enough for all characteristics in order to obtain high scores. For more information see section 3.11 about the combination of all three main relations.

The number of newly classified nouns is presented in correlation to some recall values, to gain a better understanding of the recall measure. **For example a recall of 31% results in 11,000 new nouns in the case of adjective-results and about 5,700 new nouns for verbal relations.** In case of a recall around 60%, more than 50,000 new nouns get a classification.

3.11. Combination of Modifying and Verbal Relations

After having examined the three major groups and theta-roles with equal settings and finding optimal parameter combinations, one way to improve solutions is by combining the individual results of each relation. This was done for the three major groups named *adjective, object and subject*. The three relations were bootstrapped separately. The results were then listed and compared. Then only those results which were equal in all three relations are extracted and used for composing complex semantic classes. Furthermore words whose values were equal twice were analyzed. So if one word has been assigned a positive value for one characteristic twice, it gets this value in the combination.

First binary features are analyzed and then their combination.

3.11.1. Binary Features and Ontological Sorts

Precision of the combination is almost exclusively above 99% for features and ontological sorts for most parameter settings. Even precision for smaller classes is high, but sometimes with very low recall.

One discovery is that even with the parameters

- **maxClass set to 2**, which allows a bootword to co-occur with two classes and
- **minBoot set to 1** so one single noun of a class is enough

Chapter 3. Experiments, Results and Interpretation

precision of their combination is still in the high nineties. This is because the combination of three relations makes these parameters redundant. If only values are used that were assigned three times, then minBoot is automatically not only one but three, only across different relations.

With the results of section 3.11.2 in mind, table 3.13 in this section and B.8 in the appendix list all important results of the combination. Precision values are listed for results that were the same three times and for results that were the same twice. Table 3.13 shows results for the parameter combination 2-1. One noun of a class that co-occurs with a bootword is enough, even if this bootword co-occurs with a noun of a different class. This is the minimal setting which poses the least restrictions on recall. Precision (abbreviated with 'Prec.' in the table) is listed for adjectives, objects and subjects in the first three columns. The corresponding recall values follow. The subsequent columns show the results of the combination. The first lists precision values of the combination where the results of the three relations are all equal. Notice how these are at 99% in many cases. Only a few characteristics show less then 99% with the smallest value at 93%. The precision of the smaller positive ('pos.') classes is also above 70% in many cases, which is better than in previous experiments. Precision of the combination with two equal values from the three relations is still very high with only few cases below 90%. Furthermore the recall is about 20% higher for these combinations as seen in the last two columns.

Characteristic	Prec. adj	Prec. obj	Prec. subj	Recall adj	Recall obj	Recall subj	Prec. 3Same	Prec. 3Same pos.	Prec. 2Same	Prec. 2Same pos.	Recall 3Same	Recall 2Same
human:+	0.8752	0.9167	0.9271	0.6178	0.6902	0.6695	0.9932	0.9700	0.9587	0.8654	0.4447	0.6895
geogr:-	0.8498	0.8902	0.8282	0.6533	0.7118	0.4759	0.9802	0.2535	0.9275	0.1436	0.3532	0.6620
spatial:+	0.8254	0.9381	0.9251	0.5583	0.6972	0.6781	0.9942	0.9961	0.9618	0.9693	0.4002	0.6864
legper:+	0.8721	0.9176	0.9318	0.6128	0.6903	0.6732	0.9937	0.9712	0.9606	0.8708	0.4397	0.6916
instit:-	0.9319	0.9233	0.9379	0.7371	0.7483	0.7359	0.9961	0.3846	0.9741	0.0476	0.5985	0.7581
animal:-	0.9461	0.9485	0.9286	0.7478	0.7655	0.7249	0.9899	0.8372	0.9654	0.6102	0.6162	0.7589
potag:+	0.8518	0.9119	0.9519	0.5889	0.6823	0.6899	0.9933	0.9955	0.9593	0.9727	0.4267	0.6898
movable:+	0.8034	0.8724	0.8557	0.5485	0.6494	0.4381	0.9622	0.9571	0.9063	0.8938	0.2592	0.5771
animate:+	0.8643	0.9177	0.9432	0.6049	0.6873	0.6823	0.9917	0.9929	0.9582	0.9618	0.4396	0.6901
info:-	0.8256	0.8339	0.7875	0.6425	0.6705	0.6174	0.9670	0.0769	0.8723	0.0517	0.4541	0.6607
thconc:-	0.7422	0.8015	0.7853	0.5571	0.6432	0.6141	0.9320	0.3034	0.8330	0.2344	0.3879	0.6277
method:-	0.9403	0.9450	0.8953	0.7494	0.7693	0.7088	0.9954	0.0000	0.9719	0.0090	0.6089	0.7568
axial:+	0.8185	0.8720	0.8774	0.5581	0.6488	0.6395	0.9740	0.9836	0.9130	0.9232	0.3774	0.6449
mental:-	0.9085	0.8889	0.8767	0.7207	0.7207	0.6916	0.9827	0.1739	0.9366	0.1044	0.5531	0.7271
instru:-	0.8495	0.8677	0.8447	0.6284	0.6705	0.6410	0.9751	0.7814	0.9077	0.6071	0.4446	0.6692
artif:-	0.7136	0.7774	0.7766	0.4887	0.5871	0.5768	0.9210	0.7769	0.8263	0.6554	0.2930	0.5724
sort:d+	0.8155	0.9115	0.8993	0.5514	0.6745	0.6552	0.9876	0.9866	0.9409	0.9424	0.3797	0.6653
sort:na-	0.9531	0.9538	0.9549	0.7600	0.7752	0.7558	0.9791	1.0000	0.9661	0.6333	0.6554	0.7746
sort:abs-	0.7865	0.8900	0.8803	0.5503	0.6774	0.6642	0.9682	0.8940	0.9123	0.7950	0.4028	0.6628
sort:mo-	0.9594	0.9622	0.9413	0.7683	0.7833	0.7455	0.9985	0.5000	0.9861	0.0000	0.6457	0.7775
sort:ta-	0.9398	0.9314	0.8865	0.7425	0.7523	0.6989	0.9955	0.3333	0.9689	0.1068	0.5800	0.7514
sort:co+	0.8254	0.9381	0.9251	0.5583	0.6972	0.6781	0.9942	0.9961	0.9618	0.9693	0.4002	0.6864
sort:ab-	0.8293	0.9384	0.9245	0.5617	0.6983	0.6806	0.9929	0.9914	0.9645	0.9574	0.4023	0.6900
sort:s-	0.9355	0.9216	0.8835	0.7324	0.7390	0.6905	0.9921	0.5238	0.9605	0.3098	0.5628	0.7398
sort:oa-	0.9535	0.9407	0.9240	0.7616	0.7648	0.7308	0.9960	0.0000	0.9788	0.0190	0.6209	0.7682
sort:io-	0.6867	0.7809	0.7591	0.5012	0.6163	0.5835	0.9113	0.3785	0.8066	0.2814	0.3241	0.5955
sort:o+	0.9761	0.9866	0.9775	0.7814	0.8034	0.7742	0.9996	0.9996	0.9973	0.9989	0.6723	0.8004
sort:me-	0.9756	0.9866	0.9771	0.7812	0.8034	0.7740	0.9996	1.0000	0.9974	0.6286	0.6724	0.8000
sort:qn-	0.9756	0.9866	0.9771	0.7812	0.8034	0.7740	0.9996	1.0000	0.9974	0.6286	0.6724	0.8000
sort:ad-	0.7831	0.8642	0.8551	0.5572	0.6747	0.6602	0.9590	0.8351	0.8911	0.7230	0.4047	0.6595
sort:at-	0.9469	0.9461	0.9425	0.7521	0.7678	0.7451	0.9786	0.5000	0.9629	0.5507	0.6371	0.7679
sort:re-	0.9657	1.0000	1.0000	0.7727	0.8143	0.7921	1.0000	1.0000	1.0000	1.0000	0.6855	0.8040
sort:as-	0.8130	0.7976	0.7739	0.6266	0.6420	0.6084	0.9368	0.2164	0.8364	0.1461	0.4272	0.6403

Table 3.13.: Combining results of the three main relations, parameter combination: 2-1

Table B.8 shows similar results for the standard parameters 2-5, but with a slightly higher precision and lower recall.

60

Chapter 3. Experiments, Results and Interpretation

Table 3.14 displays the results of the combination based on the parameter setting 1-2, which returns the best results for single relations. All precision values are at 99% for this combination. This comes with an extremely low recall of only 58 % for the best cases but only 6 % for *spatial* or *movable*.

Characteristic	Prec. 3Same	Prec. 3Same pos.	Prec. 2Same	Prec. 2Same pos.	Recall 3Same	Recall 3Same pos.	Recall 2Same	Recall 2Same pos.
human:+	1.0000	1.0000	0.9991	1.0000	0.1546	0.0020	0.3123	0.0201
geogr:-	0.9960	1.0000	0.9902	1.0000	0.2915	0.0000	0.4742	0.0000
spatial:+	1.0000	1.0000	0.9981	0.9966	0.0620	0.0335	0.2356	0.1335
legper:+	1.0000	1.0000	0.9988	1.0000	0.1344	0.0035	0.2894	0.0281
instit:-	0.9972	1.0000	0.9947	1.0000	0.4441	0.0000	0.5963	0.0000
animal:-	0.9970	1.0000	0.9907	1.0000	0.3951	0.0000	0.5537	0.0000
potag:+	1.0000	1.0000	0.9975	0.9966	0.0977	0.0070	0.2558	0.0526
movable:+	0.9986	1.0000	0.9933	0.9858	0.0613	0.0067	0.1876	0.0439
animate:+	1.0000	1.0000	0.9984	0.9973	0.1288	0.0036	0.2871	0.0330
info:-	0.9958	1.0000	0.9895	1.0000	0.2995	0.0000	0.4732	0.0000
thconc:-	0.9814	1.0000	0.9688	1.0000	0.1901	0.0000	0.3721	0.0000
method:-	0.9981	1.0000	0.9979	1.0000	0.4855	0.0000	0.6335	0.0000
axial:+	1.0000	1.0000	0.9927	0.9966	0.0646	0.0089	0.1955	0.0521
mental:-	0.9905	1.0000	0.9855	1.0000	0.3645	0.0000	0.5201	0.0000
instru:-	0.9921	1.0000	0.9825	1.0000	0.1805	0.0000	0.3386	0.0003
artif:-	0.9909	0.5000	0.9751	0.9412	0.0393	0.0001	0.1448	0.0014
sort:d+	1.0000	1.0000	0.9986	0.9965	0.0527	0.0172	0.1950	0.0775
sort:na-	0.9945	1.0000	0.9892	1.0000	0.4230	0.0000	0.5675	0.0000
sort:abs-	0.9991	1.0000	0.9967	1.0000	0.1026	0.0014	0.2752	0.0141
sort:mo-	0.9985	1.0000	0.9988	1.0000	0.5381	0.0000	0.6680	0.0000
sort:ta-	0.9972	1.0000	0.9938	1.0000	0.4100	0.0000	0.5752	0.0000
sort:co+	1.0000	1.0000	0.9981	0.9966	0.0620	0.0335	0.2356	0.1335
sort:ab-	1.0000	1.0000	0.9996	1.0000	0.0623	0.0193	0.2399	0.0787
sort:s-	0.9965	1.0000	0.9908	1.0000	0.3597	0.0000	0.5263	0.0000
sort:on-	0.9972	1.0000	0.9968	1.0000	0.4878	0.0000	0.6372	0.0000
sort:io-	0.9894	1.0000	0.9704	1.0000	0.1341	0.0000	0.3071	0.0000
sort:o+	1.0000	1.0000	0.9995	0.9995	0.5538	0.5537	0.6801	0.6801
sort:me-	1.0000	1.0000	0.9995	1.0000	0.5534	0.0000	0.6804	0.0000
sort:qn-	1.0000	1.0000	0.9995	1.0000	0.5534	0.0000	0.6804	0.0000
sort:ad-	0.9970	1.0000	0.9884	1.0000	0.1202	0.0002	0.2830	0.0011
sort:at-	0.9945	1.0000	0.9899	1.0000	0.3884	0.0000	0.5402	0.0000
sort:re-	1.0000	1.0000	1.0000	1.0000	0.5779	0.0000	0.6977	0.0000
sort:as-	0.9880	1.0000	0.9768	1.0000	0.2303	0.0000	0.4053	0.0000

Table 3.14.: Combining results of the three main relations with parameters in single runs: maxClass=2 and minBoot=1

These high quality results of binary characteristic with the 1-2 parameter combination, cannot be used in the combination to semantic sorts, since the recall for a few characteristics was not high enough. These characteristics were hence underspecified and so the corresponding noun could not be assigned a semantic sort. In environments where an incomplete and high quality set of features is possible, this parameter combination is the best choice.

3.11.2. Composing Semantic Sorts with Combined Results

The improved results of binary features can successfully be employed to compose complex semantic sorts with higher precision than for each separate relation. As seen before there is a trade-off between high precision and recall. The combination 2-1 results in the highest recall of 26% which is only as high as for subjects, so subjects shall be used instead. The highest precision based on the Wikipedia corpus was 78.3% and achieved with the parameters 2-5 and the combination of all three relations. **This is an improvement of 2.5% over the best combination from a single relation.** With

Chapter 3. Experiments, Results and Interpretation

a recall of 10.5% about 1200 of the known nouns got classified. Table 3.15 shows these results.

Sort	Total Distribution	Correct + False	Correct	% Recall	% Precision
nonment-dyn-abs-situation	2751	24.79	530	448	16.28	84.53
human-object	2359	21.26	602	586	24.84	97.34
prot-theor-concept	815	7.34	71	54	6.63	76.06
nonoper-attribute	635	5.72	0	0	0.00	0.00
animal-object	593	5.34	14	14	2.36	100.00
ax-mov-art-discrete	568	5.12	6	2	0.35	33.33
plant-object	445	4.01	7	7	1.57	100.00
nonment-stat-abs-situation	378	3.41	12	4	1.06	33.33
nonmov-art-discrete	191	1.72	37	14	7.33	37.84
nonax-mov-art-discrete	174	1.57	17	1	0.57	5.88
mov-nonanimate-con-potag	159	1.43	6	6	3.77	100.00
ment-stat-abs-situation	156	1.41	1	0	0.00	0.00
abs-info	148	1.33	10	7	4.73	70.00
art-substance	147	1.32	3	1	0.68	33.33
nat-discrete	146	1.32	2	0	0.00	0.00
tem-abstractum	143	1.29	6	4	2.80	66.67
art-con-geogr	138	1.24	6	5	3.62	83.33
prot-discrete	135	1.22	35	2	1.48	5.71
nat-substance	130	1.17	3	2	1.54	66.67
nat-con-geogr	105	0.95	1	1	0.95	100.00

Table 3.15.: Semantic sorts composed from the combination of three equal values from the three main relations, which were bootstrapped separately with the parameters 2-5

If two equal values are sufficient for assigning a class, recall is at 22.4% with a precision of 76.3% for the same parameter combination. This way 2486 new correct nouns are found and 3258 total new nouns are found.

As observed during experiments with the PDW in section 3.7.3, these values improve significantly, if the same method is applied on a bigger, high quality corpus.

This section contains the culmination of the experiments and precision values. It shows that **binary features can reach 99% precision if the three main relations are combined, compared and used only if equal.** These binary features also resulted in a 2.5% gain for the combination to complex semantic classes, over previous separate combinations.

Chapter 3. Experiments, Results and Interpretation

3.12. Error Analysis

One thing that comes to attention when looking at all results is that the feature *artif*, (i.e. artifact, such as house) is constantly receiving a precision below average. This is due to the circumstance that most verbs do not speak clearly for or against a noun to be an artefact, because most actions that can be performed on artefacts can also be performed on natural objects (i.e. pour gas(+) and water(-), spend freetime(+) and time(-), sit computer(+) and meadow(-), etc.).

The main reason for most semantic sorts to have a zero precision is that they do not occur as a subject or as the first complement of a verb. However for a few this is not the case. The following three sorts do appear, but very rarely get specified correctly. One reason might be that a few single unspecified characteristics with low recall result in ambiguous sorts since they distinguish one sort from another. Finding these automatically is one possible improvement in future setups. Other reasons are listed below.

For *nonoper-attribute* (attributes that cannot be measured, such as gentleness, futility, neutrality) and *ment-stat-abs-situation* (such as frustration, paranoia, passion) one problem is that they occur mostly with unspecific adjectives such as *high, big, small, fine*, etc. If one of these occurs as an object, the verbs are also very general and could be applied to an opposite class. Take the verb-object pair *injure-neutrality*, another pair might as well be *injure-nose*, which would then be of sort *nat-discrete*. Also often these appear with verbs such as *have, get, become, be*, which prevent successful classification.

nat-discrete (any natural discrete objects, liver, blossom, beard) on the other hand appear mostly with color adjectives, which themselves can modify artefacts as well. That is why precision for adjectives is low. For objects the bootstrapping process is confused, when a blossom is manufactured (a plastic one). Most people *have or carry* a liver, which is not supporting the classification.

One problem however is not to be found in the algorithm or the corpus, but in an uncomplete entry for the German word 'Blüte', which is not differentiated in meaning as it should be. Thereby, the same meaning occurs falsely where it should have been contextually disambiguated by the parser into one of its meanings, take for example these phrases that all use the German wird (Blüte).

- grunge reached its full flowering (Blüte) = Grunge erreichte seine Blüte
- the flower of the young woman faded = die Blüte der Jugend der jungen Frau welkte
- the hyacinth blossom (Blüte) = die Hyazhinten Blüte
- the dud banknote was found = die Blüte wurde gefunden

Uncomplete entries are found in this approach and could be automatically extracted for future lexicon and parser improvements.

Chapter 3. Experiments, Results and Interpretation

Section 4.1 shows four possible improvements of the combination method.

3.13. Summary

This chapter describes the results of all experiments in detail. First the three corpora on which the bootstrapping was carried out were introduced. After explaining all the different measures used to analyze the results, the three main groups were examined. These are the modifying relation between adjectives and nouns, and the verbal relations between:

1. a verb and its subject noun (first complement in semantic nets)
2. a verb and its object noun (second complement in semantic nets)
3. a verb and its theta-roles (saved in verb valency lists)

During the experiments it became evident that verbal relations, especially verb-object relations, return the highest precision values for binary characteristics. Subjects however provide higher precision for the combination to complex semantic classes. For verification the same experiments were carried out with a different corpus, which provided the same relative results, with regard to verbal relations. However average precision for the combination to semantic classes was significantly higher. This lead to the hypothesis that a high quality linguistically balanced corpus is more useful for the final combination to HaGenLex' semantic sorts than a corpus like Wikipedia could be.

Furthermore it was found out that theta-roles were useful for certain characteristics. The reason why they could not be used exclusively was a smaller number of extracted pairs.

Different parameter settings were compared. This demonstrated that both smoothing and the extraction of only significant pairs are both not improving the results tremendously. In environments such as the one presented in this work the 1% gain is still important though and thus both are going to be used in the future.

One new and interesting combination of parameters lead to the conclusion that it is more useful to use fewer (e.g. three) strict bootwords which co-occur with one class exclusively than more (e.g. five) which can occur with different classes. This finding could be used to make the algorithm more effective, since no probabilities have to be calculated. However since recall drops enormously for some characteristics, these results can only be used in environments where a subset of all possible values is sufficient.

The culmination of the presented experiments is the combination of the results of the three main relations which showed significantly better results than single relations. Most binary characteristics gained precision values of 99%. The combination to semantic sorts also showed improvements of about 2.5%, resulting in an overall precision of 78.3%, which

Chapter 3. Experiments, Results and Interpretation

can be improved by a larger high quality corpus as shown with a small subset of the PDW corpus.

Chapter 4.
Improvements and a Novel Approach

This chapter introduces several improvements of this method. Some are based on the used lexicon, others are generally applicable on other lexicons as well. Furthermore suggestions for practical use and an absolutely novel approach of genetic meta-bootstrapping are provided.

4.1. Combination of Characteristics to Complex Semantic Sorts

Any lexicon that defines complex semantic sorts or classes can be extended with the presented method. However the used method is not yet optimal.

While results of binary characteristics show much improvement in verbal relations, the overall performance for their combination to complex semantic sorts is not satisfying. Improving the algorithm for combining them after the bootstrapping process is the goal of this section.

In order to prove the general hypothesis that verbal relations return better results, experiments with a small subset of the *Projekt Deutscher Wortschatz* were carried out. Here precision was higher for subject-verb relations than in any other previous study and higher than in experiments based on the Wikipedia and CLEF corpus. This demonstrates that a high quality corpus is very important for the general bootstrapping outcome. Since ratios of bootword over noun in the bipartite graph were different, the hypothesis that higher results of verbal relations in other experiments were merely based on these rations could be proven as unjustified.

However the combination method bears many possibilities of improvement. First of all a detailed debug-mode for wrong combinations could help finding specific characteristics which prevent certain sorts to be combined successfully. Then these could be assigned a default in some cases.

In a second step, instead of not assigning a class at all, if there is more than one element missing, the most general of the possible classes could be assigned. In a third modification, uncertain characteristics (based on tests before, or the *doubt* measure) could be ignored and set to a default value, if this value is the only missing part of a semantic sort.

The fourth improvement could exploit the strict hierarchy of sorts and the loose combinations of features. Take for example the feature *artif* (artefact), which shows very low precision. In MultiNet the following rule is valid.

$$\forall x[artif(x) \rightarrow \neg animal(x)]$$

The functions named after features return true if and only if, the parameter is positive for this feature. If for example a specific x is animal:+ (i.e. $\neg animal(x)$ is false) with a high probability, it could be inferred with the modus tollens that:

$$\frac{artif(x) \rightarrow \neg animal(x), animal(x)}{\neg artif(x)}$$

This way the often correctly identified feature *animal* can be used to define the feature *artif*. These information are saved electronically at the Fernuniversität Hagen and could be used for this method.

Last but not least the best sorts can be chosen among runs from different relations.

4.2. Selection Procedure between Runs

Another improvement could use the different relations that were separately bootstrapped. Each class is saved with its probability and afterwards all results are compared and only the most probable pieces are taken.

It is important to notice that these procedures should be separated in the beginning. The combination of adjective-, subject and object-relations into one graph has not shown better results than each of them alone.

Similarly to section 4.3 and quite easy to apply could be the rule that only these characteristics are used that were the same in alle different runs.

4.3. Using the Doubt Measure

A new measure that has been introduced in this work is *doubt*. It shows how often a noun gets a different class between runs during the 10-fold-cross-validation method, which is normally not used if the algorithm is applied in practice. However since *doubt* is calculated only on the basis of unknown nouns, it might as well be used for extending the lexicon. All seed words can be partitioned into n-sets. Then each set is bootstrapped with the help of a corpus. All results are combined. Then only these nouns are used for further processing that have had the same class in all runs. This way the solution is more robust and less dependent on the seed words.

To better understand the idea a few example values are given in section 3.7.4.

Furthermore a correlation between doubt and precision could be established by creating overlapping test sets and comparing the outcomes of different runs with different sub-samples of the corpus. Afterwards the established correlation can be used for sorting out unknown nouns, which show a doubt value above a certain threshold. Further investigating this measure and experimenting with correlation values shall be performed in future works.

4.4. Genetic Bootstrapping

The presented approach has an inherent problem. Specific contexts (or extraction patterns) such as adjectives, or subjects cannot bootstrap all noun classes, since some of them do not occur in this context. They can be combined afterwards, but might still have the same problem.

Thus, a novel approach to bootstrapping is proposed here, which combines ideas of genetic algorithms and bootstrapping, in order to improve precision and increase recall.

4.4.1. Genetic Algorithms

This machine learning theory tries to solve problems by a process of evolution to the best solution similar to evolutionary process in nature. Each run of a genetic algorithm starts with a population of individuals, represented by chromosomes. In each iteration chromosomes can be exchanged between individuals, which is called crossover. Whose chromosomes can be passed on to many other individuals is defined by the fitness functions. Before the process started the programmer has to define this function. It expresses the proximity of an individual (i.e. the solution it represents) to the solution. By means of selection of the best individuals (and sometimes mutation of the chromosomes) each generation gets closer to an optimal solution.

Each iteration is usually performed with the following steps. First the fitness value of all individuals is calculated. Then a new population is created by using crossover, mutation and (in case of great fitness) simple reproduction. Afterwards different methods can be applied to decimate this generation. This process is repeated until a threshold of fitness is reached.

The first time evolutionary computing was mentioned in I. Rechenberg's work 'Evolutionsstrategie'. However the idea fully developed only under John Holland into what is now

Chapter 4. Improvements and a Novel Approach

called genetic algorithms, he published his ideas in the book 'Adaption in Natural and Artificial Systems'in 1975 ([Gol89]).

4.4.2. Genetic Algorithms in NLP

Genetic (or evolutionary) algorithms have been applied in different fields of natural language processing ([DD96]). In [SW95] a grammar is generated with a genetic algorithm. It uses frequent sentence examples to learn patterns. However infrequent but correct tree structures also persist, once learned. This idea is picked up and developed further by [AKM03]. There, the length of the grammar in Chomsky Normal Form is included into the fitness function. Smaller grammars result in a better fitness. Negative examples, that is wrong sentences still being parsed, also discount the overall fitness.

4.4.3. The Novel Approach

The idea is that the algorithm itself finds the context, which describes a certain class the best way. This way the *Distributional Hypothesis* can be used in a greater variety and it can be discovered, whether local or global contexts play a bigger role for some classes. To my best knowledge, this approach has not been developed. It is called genetic bootstrapping, because the genetic algorithm find the best bootstrapping rule.

Data

The needed data can vary. One possibility is to use a tree bank. This way, certain sub tress can be used as chromosomes, their branches can either be defined by the included words or categories. Another simpler method is a flat only POS-tagged corpus. Other types of corpora like tree banks are possible too, the only importance is that parts of each sentence can be extracted clearly. In this short outline, a flat POS-tagged corpus shall be used.

Besides the corpus, a semantic taxonomy, or lexicon is needed, which provides the seed words and their semantic features or classes.

Definitions

First the term '**individual**' needs to clear. An individual in this case is a bootstrapping rule (also called extraction pattern) for one specific class [1].

---

[1] It could also be a set of rules for different classes, this could be created, if the method works for one class.

All rules are made of **chromosomes**, which are parts of this rule. For example one part could be that the word, whose class is to be bootstrapped, needs to be in front of the preposition 'to'. Other parts can be that it has to be followed by the category 'N' (noun). These are nearest neighbour parts. However the whole sentence can be used. One class could need the verb 'wear' inside a sentence. Depending on the level of POS-tag details, it could also be desired to have an infinite V_0 (verb), where the phrase has another verb V_0 'want'. This is just to show the many dependencies that could be discovered.

The **fitness function** is defined by precision, recall and length of a rule R. It could be calculated similar to the F-score, but including the number of parts $p(R)$. α is the weight for precision. β defines the importance of the rule length. The higher it is, the less important the rule length will be. Possible values are in the range of 0 to 1. If a certain number of parts for a rule is considered perfect, this number could the value of the threshold t, which determines whether, the length play a role or not.

$$F_{\alpha,\beta}(R) = \left[\frac{(1+\alpha) \times \text{precision}(R) \times \text{recall}(R)}{\alpha \times \text{precision}(R) + \text{recall}(R)}\right] \times \beta^{\max(0, p(R)-t)}$$

Mutation can delete a part of a rule, add one or replace it by another. Fragments from sentences, in which a noun of the looked for class was not found correctly in the training phrase, are more likely to contribute a mutation part. **Crossover** takes the minimum number of parts from both parents. It then selects one position randomly below this minimum and changes the parts in this position in both parents. If a duplicate rule became member of the child, one of them is deleted.

For the **initial set** of rules n sentences, with a noun of the searched for class, are taken. All possible parts are extracted from those sentences. Possible parts might be: all categories that occur m-words away from the noun, all nearest neighbour words (or q words away), the verb of the phrase where the noun occurs, etc. n does not need to be too large, a number below 15 should suffice.

The number of rules is r. Before the process starts a minimum fitness (or precision) f needs to be defined. As soon as one rule is sufficiently fit, the process stops and returns this rule. In a final run this rule bootstraps on the whole corpus and extracts other nouns of the class.

The Algorithm

Figure 4.1 describes the algorithm in pseudo code. Calculating the fitness of rules means to use them for bootstrapping the desired class on a training corpus (similar to the one described in section 2.3.2), let them find new nouns, calculate precision and recall with known nouns and combine both measures with length of the rule into fitness. For successful mutation, h sentences, which contain a known noun, are manually extracted from the corpus. This set is called 'help set'.

Chapter 4. Improvements and a Novel Approach

```
Get all parts from first training set of n sentences;
Distribute these parts evenly on all r rules;
Save h sentences with known noun of the class;

while (best rules fitness < f){
    for each(rule in set of r rules){
        calculate fitness;
    }
    crossover neighbouring rule pairs in a list sorted by fitness;
    delete the last half of all parents;
    create parts from one random element in help set;
    mutate last fourth with these parts;
}
bootstrap with best rule;
```

Figure 4.1.: Algorithm for genetic bootstrapping

Possible Drawbacks

The outlined method is quite complex since the bootstrapping process itself is time-consuming. However if implemented effectively and with a small set of rules in the beginning, it might be manageable.

Evaluations and tests of this approach shall be performed in the future, but go beyond the scope of this thesis.

Chapter 5.
Conclusion

This thesis explores a statistical bootstrapping approach for learning semantic characteristics for nouns. In particular novel verbal relations have been explored. These outperform previous experiments that bootstrapped solely on adjective-noun co-occurrences. Relations between verbs and their nominal object show especially high quality results with precision above 95% and recall at about 50%. If outcomes of bootstrapping on different relations are combined by only using nouns which were assigned the same class in the separate runs, the results can be seen as almost certain, with a precision of 99% for all binary semantic characteristics. This way recall values of the different characteristics vary between 20% and 40%. Thus some semantic characteristics can be assigned for up to 50,000 new nouns.

In the case of the lexicon in this work, characteristics are seen as binary features and ontological sorts. Together they are combined to complex semantic sorts. For these a maximum precision of 86% together with a recall of 10% are obtained with a small subset of the *Wortschatz* corpus of the University Leipzig and extracted noun-subject relations. Results with the Wikipedia corpus gain lower precision which shows a dependency on a high quality corpus. The excellent values of binary characteristics cannot be used successfully for the combination to semantic sorts because some of them receive such a low recall that they are underspecified for most nouns. These nouns are then not assigned a semantic sort.

Through extensive automatic analysis it is demonstrated that the fuzzy concepts of *subject* and *object* produce high quality results in verb-noun relations. This outcome proofs the fact that verbal relations can be successfully employed in such bootstrapping methods. It is also shown that theta-roles, that are extracted from semantically parsed corpora, can be incorporated successfully into the bootstrapping algorithm to find certain semantic characteristics with high precision.

The presented methods are applied to German word co-occurrences and are based on semantically parsed corpora. Nevertheless the core bootstrapping method can be applied to other languages and relations as well and does not necessarily need a deep semantic analysis. Through additional experiments on the balanced *Wortschatz* corpus

Chapter 5. Conclusion

uncertainty whether previous results were only caused by skewed distributions in the co-occurrence graph could be dismissed.

Not only was it discovered that verbal relations return better results, but also new and different parameter settings were systematically analyzed and new combinations that provide higher precision were found. If parameters were set in a way that strict bootwords are allowed to co-occur only with nouns of one class at least two times, precision of all characteristics is exclusively above 99 % for verb-subject relations. However recall values of different characteristics vary greatly between 8,000 and 55,000 newly identified nouns, similar to the combination of different relations. In environments where underspecified entries for some semantic characteristics are allowed, the presented method can be used to extend the lexicon with several characteristics very effectively.

During the practical part of this thesis many different applications from the University Leipzig and the Fernuniversität Hagen were combined and in some cases modified. The process from corpus extraction and preparation over pre-processing and bootstrapping was automated and parameterized. Furthermore software was written for the extensive testing and analysis of the results. Finally software was written to combine the results of different bootstrapping runs. Testing other relations for their applicability to this method should be easy with the developed tools.

In the end five improvements for the method of combining complex semantic sorts are given and an absolutely novel approach which combines ideas from genetic algorithms and bootstrapping is outlined.

With the help of detailed and systematic analyses of the results of bootstrapping on adjective-noun and verbal relations, the latter showed better overall results. These can be improved by combining both context types, which results in 99% precision for binary semantic characteristics. Thus, this thesis proposes to follow along this line and to combine bootstrapping results of different verbal and modifying relations in which nouns are defined by binary characteristics. In cases were only one relation is available new parameter combinations that demand stricter class membership, but fewer bootwords, have proven to be optimal. With the help of the presented conclusions the process of extending semantic lexicons will be more effective.

Glossary

B

bias Used for binary classifiers to show the ratio of the bigger class. Calculated by counting the number of occurrences of both classes and dividing the bigger number by the sum of both. Sometimes also used to show the ratio of the smaller class, though not in this work.

bigram In the context of this work a bigram is a group of two words.

bootstrapping In this work bootstrapping is a method by which new information about unknown words can be obtained by means of already known seed words and similar contexts of both in a corpus.

bootword Throughout this work, a bootword is used to describe the word with which a noun co-occurs. The following **bootword types** are treated: verb (bootword)-noun (object), verb-noun (subject), adjective (bootword)-noun and various relations of the type: verb - noun(with specific theta-role), p. 1.

C

classifier The term classifier refers to both features and ontological sorts.

clustering Clustering is the action of grouping together words, based on a specific similarity.

co-occurrence The above-chance frequent occurrence of two terms from a corpus next to each other in a specific order. Unlike collocation, co-occurrence assumes that the two terms are interdependent.

collocation A sequence of words or terms which co-occur more often than would be expected by chance, such as *strong tea*, or *make up*.

concordance List of all contexts of a given word in a corpus.

GLOSSARY

context In this framework, context refers to the surroundings of a word in a sentence. These can be nearest neighbour words, such as the word left and right of the given word, or other important words of the sentence such as the main verb, if the given word stands in some relations to it.

corpus A collection of text, now usually in machine-readable form and compiled to be representative of a particular kind of language and provided with some kind of annotation.For important corpus features see section 1.2.3.

D

doubt A ratio for binary classifiers that reflects the doubt of classifications by comparing the results from different training sets. Calculated for each word with the formula: $doubt = \frac{\min(class\ for\ this\ noun)}{\max(class\ for\ this\ noun)}$.

F

f-score The f-score is a measure that combines precision and recall, it is the weighted harmonic mean of both. Depending of the parameter α one value can be weighted. With an α of 0.5 precision is weighted. The f-score is calculated with the following formula: $F_\alpha = \frac{(1+\alpha) \times precision \times recall}{\alpha \times precision + recall}$.

H

hypergraph A hypergraph is a graph whose edges can branch out and connect more than two nodes.

L

lexical semantic Lexical semantic is a subfield of semantic theory, which describes word meanings in their interrelations between with each other.

P

part-of-speech (POS) The part-of-speech tag defines the category of a word, for example verb, noun, adjective.

precision The ratio of found relevant (correct) tokens over the total number of found tokes (correct and wrong). $precision = \frac{number\ of\ relevant\ found\ tokens}{total\ number\ of\ found\ tokens}$.

GLOSSARY

R

recall The ratio of found relevant tokens (correct ones) over the total number of tokens in the search domain. $recall = \frac{number\ of\ relevant\ found\ tokens}{total\ number\ of\ tokens}$.

S

seed words In this work seed words are known nouns. These nouns and their corresponding features and ontological sorts are used during the learning phase, to calculate noun profiles for each bootword. With their help new nouns can be identified by the bootstrapping algorithm.

T

theta role Theta roles are relations between a verb (or adjective or noun) and nouns in a given sentence. In this work, only theta-roles are used that are complements of the verb. Complements are nouns that fill a spot in the valency list of a verb. See sections 1.4 for more details.

token An individual word.

V

valency list Each verb in the lexicon has a valency list. In MultiNet there is a subcategorization frame which defines language dependent syntactic information for a verb. In addition there is also a language independent valency list. It defines in which theta roles (relations) a verb must stand with other elements of a sentence. These positions have to be filled by elements in order to complete the verb. However not necessarily in the syntactic surface structure of the sentence.

verbal relations This term refers to all types of relation, a verb can have with its complements or adjuncts. In this work these are subjects, objects and several theta-roles.

Appendix A.
HaGenLex Tables

Name	Meaning	Examples +	Examples −
ANIMAL	animal	⟨fox⟩	⟨person⟩
ANIMATE	living being	⟨tree⟩	⟨stone⟩
ARTIF	artifact	⟨house⟩	⟨tree⟩
AXIAL	object having a distinguished axis	⟨pencil⟩	⟨sphere⟩
GEOGR	geographical object	⟨the Alps⟩	⟨table⟩
HUMAN	human being	⟨woman⟩	⟨ape⟩
INFO	(carrier of) information	⟨book⟩	⟨grass⟩
INSTIT	institution	⟨UNO⟩	⟨apple⟩
INSTRU	instrument	⟨hammer⟩	⟨mountain⟩
LEGPER	juridical or natural person	⟨firm⟩	⟨animal⟩
MENTAL	mental object or situation	⟨pleasure⟩	⟨length⟩
METHOD	method	⟨procedure⟩	⟨book⟩
MOVABLE	object being movable	⟨car⟩	⟨forest⟩
POTAG	potential agent	⟨motor⟩	⟨poster⟩
SPATIAL	object having spatial extension	⟨table⟩	⟨idea⟩
THCONC	theoretical concept	⟨mathematics⟩	⟨pleasure⟩

Table A.1.: Semantic features ([Sch98])

Appendix A. HaGenLex Tables

entity [ent]		
object [o]		
concrete object [co]		
discrete object [d]		⟨house⟩, ⟨apple⟩, ⟨tiger⟩
substance [s]		⟨milk⟩, ⟨honey⟩, ⟨iron⟩
abstract object [ab]		
attribute [at]		
measurable attribute [oa]		⟨height⟩, ⟨weight⟩, ⟨length⟩
non-measurable attribute [na]		⟨form⟩, ⟨trait⟩, ⟨charm⟩
relationship [re]		⟨causality⟩, ⟨similarity⟩, ⟨synonymy⟩
ideal object [io]		⟨religion⟩, ⟨justice⟩, ⟨criterion⟩, ⟨category⟩
abstract temporal object [ta]		⟨Renaissance⟩, ⟨Easter⟩, ⟨holiday⟩
modality [mo]		⟨necessity⟩, ⟨intention⟩, ⟨permission⟩
situational object [abs]		
dynamic situational object [ad]		⟨race⟩, ⟨robbery⟩, ⟨movement⟩
static situational object [as]		⟨equilibrium⟩, ⟨sleep⟩
situation [si]		
dynamic situation [dy]		
action [da]		⟨write⟩, ⟨sing⟩, ⟨sell⟩, ⟨drive⟩
happening [dn]		⟨rain⟩, ⟨decay⟩, ⟨explode⟩
static situation [st]		⟨stand⟩, ⟨be ill⟩
situational descriptor [sd]		
time [t]		⟨yesterday⟩, ⟨Monday⟩, ⟨tomorrow⟩
location [l]		⟨here⟩, ⟨there⟩
modal situational descriptor [md]		⟨impossible⟩, ⟨necessary⟩, ⟨desirable⟩
quality [ql]		
property [p]		
total quality [tq]		⟨dead⟩, ⟨empty⟩, ⟨green⟩
gradable quality [gq]		
measurable quality [mq]		⟨small⟩, ⟨expensive⟩
non-measurable quality [nq]		⟨friendly⟩, ⟨tired⟩
relational quality [rq]		⟨inverse⟩, ⟨equivalent⟩, ⟨similar⟩
functional quality [fq]		
operational quality [oq]		⟨fourth⟩, ⟨last⟩, ⟨next⟩
associative quality [aq]		⟨chemical⟩, ⟨philosophical⟩
quantity [qn]		
quantificator [qf]		
numerical quantificator [nu]		⟨one⟩, ⟨two⟩, ⟨five⟩, ⟨hundred⟩
non-numerical quantificator [nn]		⟨all⟩, ⟨many⟩, ⟨several⟩
unit of measurement [me]		⟨kg⟩, ⟨meter⟩, ⟨mile⟩
measurement [m]		⟨three miles⟩, ⟨two hours⟩
graduator [gr]		
qualitative graduator [lg]		⟨very⟩, ⟨especially⟩, ⟨rather⟩
quantitative graduator [ng]		⟨almost⟩, ⟨nearly⟩, ⟨approximately⟩
formal entity [fe]		(meta level entities like figures and tables)

Table A.2.: Hierarchy of ontological sorts ([HHO03])

Appendix B.

Bootstrapping Results

Sort	Total	Wrong+Correct	Correct	%Recall	%Precicion
nonment-dyn-abs-situation	2751	766	545	19.81	71.15
human-object	2359	766	678	28.74	88.51
prot-theor-concept	815	143	43	5.28	30.07
nonoper-attribute	635	5	0	0.00	0.00
animal-object	593	36	24	4.05	66.67
ax-mov-art-discrete	568	79	31	5.46	39.24
plant-object	445	31	8	1.80	25.81
nonment-stat-abs-situation	378	60	6	1.59	10.00
nonmov-art-discrete	191	93	24	12.57	25.81
nonax-mov-art-discrete	174	42	10	5.75	23.81
mov-nonanimate-con-potag	159	10	6	3.77	60.00
ment-stat-abs-situation	156	2	0	0.00	0.00
abs-info	148	31	2	1.35	6.45
art-substance	147	6	3	2.04	50.00
nat-discrete	146	8	2	1.37	25.00
tem-abstractum	143	29	20	13.99	68.97
art-con-geogr	138	17	6	4.35	35.29
prot-discrete	135	22	1	0.74	4.55
nat-substance	130	24	4	3.08	16.67
nat-con-geogr	105	30	14	13.33	46.67

Table B.1.: Adjectives: Results of combination to semantic sorts

Sort	Total	Wrong+Correct	Correct	%Recall	%Precicion
nonment-dyn-abs-situation	2751	580	473	17.19	81.55
human-object	2359	1571	1411	59.81	89.82
prot-theor-concept	815	92	60	7.36	65.22
nonoper-attribute	635	1	0	0.00	0.00
animal-object	593	51	42	7.08	82.35
ax-mov-art-discrete	568	60	25	4.40	41.67
plant-object	445	28	9	2.02	32.14
nonment-stat-abs-situation	378	42	8	2.12	19.05
nonmov-art-discrete	191	51	14	7.33	27.45
nonax-mov-art-discrete	174	52	7	4.02	13.46
mov-nonanimate-con-potag	159	28	15	9.43	53.57
ment-stat-abs-situation	156	5	2	1.28	40.00
abs-info	148	40	22	14.86	55.00
art-substance	147	17	5	3.40	29.41
nat-discrete	146	4	0	0.00	0.00
tem-abstractum	143	13	8	5.59	61.54
art-con-geogr	138	23	7	5.07	30.43
prot-discrete	135	7	0	0.00	0.00
nat-substance	130	42	13	10.00	30.95
nat-con-geogr	105	19	7	6.67	36.84

Table B.2.: Subjects: Results of combination to semantic sorts

Appendix B. Bootstrapping Results

Characteristic	Number+	Number-	Bias △	Precision	Recall	Fscore
spatial	4707	4356	0.5194	0.9814	0.4704	0.7205
axial	3744	5316	0.5868	0.9020	0.4328	0.6626
movable	3679	5384	0.5941	0.9109	0.4375	0.6694
potag	2939	6118	0.6755	0.9587	0.4641	0.7074
artif	2799	6265	0.6912	0.8350	0.4100	0.6206
animate	2697	6355	0.7021	0.9634	0.4672	0.7115
legper	2088	6964	0.7693	0.9657	0.4716	0.7158
human	2028	7024	0.7760	0.9633	0.4707	0.7142
instru	1347	7712	0.8513	0.9142	0.4648	0.6914
thconc	702	8351	0.9225	0.8943	0.4812	0.6953
animal	415	8633	0.9541	0.9804	0.5321	0.7654
geogr	276	8772	0.9695	0.9525	0.5098	0.7387
mental	204	8844	0.9775	0.9479	0.5179	0.7424
info	182	8866	0.9799	0.9424	0.5087	0.7339
instit	60	8988	0.9934	0.9883	0.5395	0.7738
method	22	9027	0.9976	0.9969	0.5479	0.7830

Table B.3.: Objects: features, distribution, bias and bootstrapping results

Characteristic	Number+	Number-	Bias △	Precision	Recall	Fscore
sort:co	4707	4356	0.5194	0.9814	0.4704	0.7205
sort:d	4346	4716	0.5204	0.9516	0.4532	0.6964
sort:ab	4280	4778	0.5275	0.9790	0.4700	0.7193
sort:abs	2824	6229	0.6881	0.9356	0.4648	0.6994
sort:ad	2336	6716	0.7419	0.8969	0.4631	0.6835
sort:io	936	8117	0.8966	0.8817	0.4627	0.6773
sort:as	488	8561	0.9461	0.8625	0.4660	0.6719
sort:at	390	8658	0.9569	0.9748	0.5334	0.7641
sort:na	353	8695	0.9610	0.9791	0.5370	0.7683
sort:s	309	8740	0.9659	0.9717	0.5223	0.7551
sort:ta	118	8930	0.9870	0.9842	0.5347	0.7688
sort:o	9012	61	0.9933	0.9995	0.5495	0.7852
sort:oa	37	9011	0.9959	0.9956	0.5462	0.7813
sort:me	36	9012	0.9960	0.9995	0.5495	0.7852
sort:qn	36	9012	0.9960	0.9995	0.5495	0.7852
sort:mo	11	9037	0.9988	0.9982	0.5487	0.7841
sort:re	0	9048	1.0000	1.0000	0.5499	0.7857

Table B.4.: Objects: sorts, distribution, bias and bootstrapping results

Characteristic	Correct:+	Wrong:+	Precision
method	0	7	0.0000
instit	2	35	0.0540
mental	19	187	0.0922
info	23	280	0.0759
animal	33	10	0.7674
geogr	57	174	0.2467
thconc	218	368	0.3720
instru	492	211	0.6998
human	916	96	0.9051
legper	970	106	0.9014
animate	1112	45	0.9611
artif	1229	460	0.7276
potag	1253	55	0.9579
axial	1743	231	0.8829
movable	1800	215	0.8933
spatial	2490	41	0.9838
Mean	-	-	0.5950
sort:mo	0	2	0.0
sort:re	0	0	1.0
sort:oa	0	4	0.0
sort:na	3	2	0.6
sort:me	3	0	1.0
sort:qn	3	0	1.0
sort:ta	3	42	0.0666
sort:at	9	11	0.4500
sort:s	52	61	0.4601
sort:as	128	613	0.1727
sort:io	356	444	0.4450
sort:ad	1238	367	0.7713
sort:abs	1586	265	0.8568
sort:d	2361	133	0.9466
sort:ab	2461	39	0.9844
sort:co-	2490	51	0.9838
sort:o-	3	3	0.5
Mean	-	-	0.6112

Table B.5.: Objects: Correct classifications of smaller (positive) classes

80

Appendix B. Bootstrapping Results

Sort	Total	Wrong+Correct	Correct	%Recall	%Precicion
nonment-dyn-abs-situation	2751	961	815	29.63	84.81
human-object	2359	1079	997	42.26	92.40
prot-theor-concept	815	128	82	10.06	64.06
nonoper-attribute	635	4	1	0.16	25.00
animal-object	593	36	33	5.56	91.67
ax-mov-art-discrete	568	175	78	13.73	44.57
plant-object	445	33	16	3.60	48.48
nonment-stat-abs-situation	378	67	26	6.88	38.81
nonmov-art-discrete	191	82	21	10.99	25.61
nonax-mov-art-discrete	174	80	15	8.62	18.75
mov-nonanimate-con-potag	159	28	17	10.69	60.71
ment-stat-abs-situation	156	8	1	0.64	12.50
abs-info	148	37	27	18.24	72.97
art-substance	147	33	10	6.80	30.30
nat-discrete	146	9	2	1.37	22.22
tem-abstractum	143	29	21	14.69	72.41
art-con-geogr	138	40	17	12.32	42.50
prot-discrete	135	21	2	1.48	9.52
nat-substance	130	36	16	12.31	44.44
nat-con-geogr	105	23	2	1.90	8.70

Table B.6.: Objects: Results of combination to semantic sorts

Characteristic	Total Number	Number+	Number-	Bias △	Precision	Recall	Fscore
human	3264	2090	1174	0.6403	0.8925	0.1377	0.3156
legper	3265	2152	1113	0.6591	0.9055	0.1433	0.3266
animate	3264	2654	610	0.8131	0.9300	0.1485	0.3376
movable	3270	2688	582	0.8220	0.9279	0.1495	0.3391
artif	3265	460	2805	0.8591	0.9557	0.1574	0.3552
potag	3271	2872	399	0.8780	0.9675	0.1635	0.3666
axial	3270	2875	395	0.8792	0.9386	0.1529	0.3459
animal	3258	385	2873	0.8818	0.9673	0.1624	0.3648
sort:d	3270	2948	322	0.9015	0.9285	0.1508	0.3415
spatial	3270	3011	259	0.9208	0.9516	0.1575	0.3549
sort:co	3270	3011	259	0.9208	0.9516	0.1575	0.3549
instru	3264	234	3030	0.9283	0.9741	0.1659	0.3712
sort:ab	3259	222	3037	0.9319	0.9620	0.1595	0.3594
sort:io	3259	169	3090	0.9481	0.9567	0.1591	0.3582
theonc	3258	76	3182	0.9767	0.9979	0.1721	0.3838
instit	3259	62	3197	0.9810	0.9567	0.1591	0.3582
geogr	3258	42	3216	0.9871	0.9958	0.1721	0.3836
sort:ta	3258	29	3229	0.9911	36526	0.1730	0.3855
info	3258	23	3235	0.9929	1	0.1730	0.3855
sort:s	3258	20	3238	0.9939	1	0.1730	0.3855
sort:at	3258	17	3241	0.9948	1	0.1730	0.3855
sort:o	3271	3256	15	0.9954	0.9995	0.1729	0.3853
sort:na	3258	11	3247	0.9966	1	0.1730	0.3855
sort:abs	3258	6	3252	0.9982	1	0.1730	0.3855
sort:ad	3258	6	3252	0.9982	1	0.1730	0.3855
sort:oa	3258	6	3252	0.9982	1	0.1730	0.3855
sort:qn	3258	2	3256	0.9994	0.9995	0.1729	0.3853
sort:me	3258	2	3256	0.9994	0.9995	0.1729	0.3853
sort:mo	3258	1	3257	0.9997	1	0.1730	0.3855
sort:as	3258	0	3258	1	1	0.1730	0.3855
mental:	3258	0	3258	1	1	0.1730	0.3855
method	3258	0	3258	1	1	0.1730	0.3855
sort:re	3258	0	3258	1	1	0.1730	0.3855

Table B.7.: All bootstrapping results for theta-role *agt*

Appendix B. Bootstrapping Results

Characteristic	Prec. adj	Prec. obj	Prec. subj	Recall adj	Recall obj	Recall subj	Prec. 3Same	Prec. 3Same pos.	Prec. 2Same	Prec. 2Same pos.	Recall 3Same	Recall 2Same
human:+	0.9205	0.9633	0.9588	0.4154	0.4707	0.4233	0.9979	0.9952	0.9820	0.9492	0.2553	0.4520
geogr:-	0.9406	0.9525	0.9382	0.4825	0.5098	0.4704	0.9907	0.5000	0.9744	0.4176	0.3446	0.5010
spatial:+	0.8611	0.9814	0.9628	0.3625	0.4703	0.4357	0.9992	0.9985	0.9855	0.9858	0.2223	0.4417
legper:+	0.9154	0.9657	0.9631	0.4103	0.4716	0.4258	0.9986	0.9955	0.9834	0.9512	0.2519	0.4526
instit:-	0.9872	0.9883	0.9900	0.5287	0.5395	0.4986	0.9964	1.0000	0.9931	0.0000	0.4010	0.5297
animal:-	0.9786	0.9804	0.9732	0.5234	0.5321	0.4866	0.9925	1.0000	0.9817	1.0000	0.3949	0.5216
potag:+	0.9003	0.9587	0.9844	0.3930	0.4641	0.4370	0.9974	1.0000	0.9832	0.9900	0.2404	0.4494
movable:+	0.8434	0.9109	0.8905	0.3605	0.4374	0.3985	0.9749	0.9823	0.9308	0.9361	0.2066	0.4097
animate:+	0.9120	0.9634	0.9727	0.4062	0.4671	0.4307	0.9964	0.9980	0.9819	0.9814	0.2497	0.4497
info:-	0.9316	0.9424	0.9226	0.4871	0.5087	0.4640	0.9928	0.3333	0.9715	0.1569	0.3469	0.4968
thconc:-	0.8244	0.8945	0.8871	0.4085	0.4812	0.4446	0.9681	0.5952	0.9166	0.4092	0.2820	0.4556
method:-	0.9955	0.9969	0.9912	0.5391	0.5479	0.5064	0.9978	1.0000	0.9978	1.0000	0.4166	0.5365
axial:+	0.8510	0.9020	0.9041	0.3633	0.4328	0.4052	0.9777	0.9935	0.9330	0.9472	0.2091	0.4105
mental:-	0.9622	0.9479	0.9512	0.5176	0.5179	0.4833	0.9856	0.0000	0.9741	0.2895	0.3827	0.5123
instru:-	0.9053	0.9142	0.9133	0.4386	0.4648	0.4365	0.9817	0.8542	0.9444	0.7964	0.2853	0.4545
artif:-	0.7589	0.8352	0.8371	0.3260	0.4099	0.3867	0.9520	0.8511	0.8805	0.7676	0.1770	0.3797
sort:d+	0.8513	0.9516	0.9343	0.3584	0.4532	0.4186	0.9903	0.9880	0.9667	0.9651	0.2113	0.4268
sort:na-	0.9792	0.9791	0.9882	0.5306	0.5370	0.5045	0.9888	1.0000	0.9831	1.0000	0.4144	0.5295
sort:abs-	0.8281	0.9356	0.9218	0.3685	0.4648	0.4367	0.9842	0.9401	0.9496	0.8742	0.2359	0.4378
sort:mo-	0.9979	0.9982	0.9981	0.5441	0.5487	0.5102	0.9981	1.0000	0.9985	1.0000	0.4230	0.5387
sort:ta-	0.9862	0.9842	0.9771	0.5274	0.5347	0.4959	0.9957	1.0000	0.9914	0.3333	0.3980	0.5275
sort:co+	0.8611	0.9814	0.9628	0.3625	0.4703	0.4357	0.9992	0.9985	0.9855	0.9858	0.2223	0.4417
sort:ab-	0.8670	0.9790	0.9620	0.3659	0.4699	0.4383	0.9976	1.0000	0.9862	0.9877	0.2244	0.4446
sort:s-	0.9732	0.9717	0.9638	0.5135	0.5223	0.4824	0.9939	0.6000	0.9834	0.7222	0.3796	0.5131
sort:oa-	0.9950	0.9956	0.9917	0.5405	0.5462	0.5059	0.9963	1.0000	0.9963	1.0000	0.4167	0.5365
sort:io-	0.7706	0.8819	0.8685	0.3657	0.4627	0.4236	0.9707	0.6988	0.9049	0.4918	0.2361	0.4303
sort:o+	0.9960	0.9995	0.9988	0.5428	0.5495	0.5106	0.9996	0.9996	0.9988	0.9988	0.4236	0.5387
sort:me-	0.9960	0.9995	0.9988	0.5430	0.5495	0.5106	0.9996	1.0000	0.9988	1.0000	0.4238	0.5387
sort:qn-	0.9960	0.9995	0.9988	0.5430	0.5495	0.5106	0.9996	1.0000	0.9988	1.0000	0.4238	0.5387
sort:ad-	0.8189	0.8969	0.8858	0.3739	0.4631	0.4351	0.9732	0.8789	0.9173	0.7837	0.2421	0.4338
sort:at-	0.9754	0.9748	0.9825	0.5253	0.5334	0.5009	0.9873	1.0000	0.9804	0.7500	0.4061	0.5266
sort:re-	0.9985	1.0000	1.0000	0.5438	0.5499	0.5113	1.0000	1.0000	1.0000	1.0000	0.4236	0.5396
sort:as-	0.8778	0.8625	0.8523	0.4525	0.4660	0.4308	0.9672	0.3036	0.9032	0.2014	0.2979	0.4559

Table B.8.: Combining results of the three main relations, parameter combination: 2-5

Bibliography

[AAA05] *Proceedings of the 20th National Conference on Artificial Intelligence*, Pittsburgh, 2005.

[AG00] Eugene Agichtein and Luis Gravano. Snowball: Extracting relations from large plain-text collections. In *Proceedings of the Fifth ACM International Conference on Digital Libraries*, 2000.

[AKM03] Meg Aycinena, Mykel Kochenderfer, and David Mulford. An evolutionary approach to natural language grammar induction, 2003.

[BBH+04] Chr. Biemann, S. Bordag, G. Heyer, U. Quasthoff, and Chr. Wolff. Language-independent methods for compiling monolingual lexical data. In *Proceedings of CicLING 2004*, Seoul, Korea, 2004.

[BKW02] Bernd Bohnet, Stefan Klatt, and Leo Wanner. A bootstrapping approach to automatic annotation of functional information to adjectives with an application to german, 2002.

[BLP89] István Bátori, Winfried Lenders, and Wolfgang Putschke. *Computational Linguistics*. de Gruyter, 1989.

[BO05] C. Biemann and R. Osswald. Automatische erweiterung eines semantikbasierten lexikons durch bootstrapping auf großen korpora. In B. Schröder B. Fisseni, H.-C. Schmitz and P. Wagner, editors, *Sprachtechnologie, mobile Kommunikation und linguistische Ressourcen Beiträge zur GLDV-Tagung 2005 in Bonn*, pages 15–27, Frankfurt am Main, 2005. Peter Lang.

[Car03] John Carroll. *The Oxford Handbook of Computational Linguistics*, chapter 12 Parsing, pages 233 – 248. Oxford University Press, 2003.

[Cur04] J. Curran. *From Distributional to Semantic Similarity*. PhD thesis, Edinburgh, 2004.

[DD96] Ted E. Dunning and Mark W. Davis. Evolutionary algorithms for natural language processing. In John R. Koza, editor, *Late Breaking Papers at the Genetic Programming 1996 Conference Stanford University July 28-31, 1996*, pages 16–23, Stanford University, CA, USA, jul 1996. Stanford Bookstore.

Bibliography

[Die05] Reinhard Diestel. *Graph Theory*, volume 173 of *Graduate Texts in Mathematics*. Springer-Verlag, Heidelberg, third edition, July 2005.

[DLR77] A.P. Dempster, N.M. Laird, and Rubin. Maximum likelihood from incomplete data via the EM algorithm. volume 39(B), pages 1–38. J. Royal Statist. Soc., 1977.

[Dow91] David Dowty. Thematic proto-roles and argument selection. *Language*, 67(3):547–619, 1991.

[Dun93] Ted E. Dunning. Accurate methods for the statistics of surprise and coincidence. *Computational Linguistics*, 19(1):61–74, 1993.

[EH96] Judith Eckle and Ulrich Heid. Extracting raw material for a german subcategorization lexicon from newspaper text. *In Proceedings of the 4th International Conference on Computational Lexicography*, 1996.

[EKPP03] Katrin Erk, Andrea Kowalski, Sebastian Padó, and Manfred Pinkal. Towards a resource for lexical semantics: A large german corpus with extensive semantic annotation. *In Proceedings of ACL*, 2003.

[Eng88] U. Engel. *Deutsche Grammatik*. Julius Groos Verlag, Heidelberg, 1988.

[FBS02] C. J. Fillmore, C. F. Baker, and H. Sato. The framenet database and software tools. *In Proceedings of LREC '02*, 2002.

[Fil68] Charles J. Fillmore. *Universals in Linguistic Theory*, chapter The case for case, pages 1–90. Holt, Rinehart & Winston, New York, 1968.

[Gol89] David E. Goldberg. *Genetic Algorithms in Search, Optimization, and Machine Learning*. Addison-Wesley, 1989.

[Gru65] Jeffrey Gruber. *Studies in Lexical Relations*. PhD thesis, Indiana University Linguistics Club, 1965.

[Har68] Z. Harris. *Mathematical structures of language*. New York: Wiley Interscience., 1968.

[Har03] Sven Hartrumpf. *Hybrid Disambiguation in Natural Language Analysis*. Der Andere Verlag, Osnabrück, Germany, 2003.

[Har05] Sven Hartrumpf. Extending knowledge and deepening linguistic processing for question answering. *QA@CLEF 2005*, 2005.

[HCJ00] Helbig H., Gnörlich C., and Leveling J. *Sprachtechnologie für eine dynamische Wirtschaft im Medienzeitalter*, chapter Natürlichsprachlicher Zugang zu Informationsanbietern im Internet und zu lokalen Datenbanken. Wien, Austria, 2000.

[Hel01] Hermann Helbig. *Die semantische Struktur natürlicher Sprache*. Springer, 2001.

Bibliography

[HH97] H. Helbig and S. Hartrumpf. Word class functions for syntactic-semantic analysis. In *2nd International Conference on Recent Advances in Natural Language Processing (RANLP-97)*, pages 312–317, Tzigov Chark,Bulgaria, Sept. 1997.

[HHO03] Sven Hartrumpf, Hermann Helbig, and Rainer Osswald. The semantically based computer lexicon HaGenLex, structure and technological environment. *Traitement automatique des langues*, 44:81–105, 2003.

[HK02] Ulrich Heid and Hannah Kermes. Providing lexicographers with corpus evidence for fine-grained syntactic descriptions: Adjectives taking subject and complement clauses. In Anna Braasch and Claus Povlsen, editors, *Proceedings of the Tenth EURALEX International Congress*, volume In Proceedings of the 4th International Conference on Computational Lexicography, Denmark, Kopenhagen, 2002.

[HOT06] Masato Hagiwara, Yasuhiro Ogawa, and Katsuhiko Toyama. Selection of effective contextual information for automatic synonym acquisition. In *Proceedings of the 21st International Conference on Computational Linguistics and 44th Annual Meeting of the Association for Computational Linguistics*, pages 353–360, Sydney, Australia, July 2006. Association for Computational Linguistics.

[HQW05] G. Heyer, U. Quasthoff, and T. Wittig. *Wissensrohstoff Text; Text Mining: Konzepte, Algorithmen, Ergebnisse*. w3l-Verlag, Bochum, 2005.

[iWSR+01] Sabine Schulte im Walde, Helmut Schmid, Mats Rooth, Stefan Riezler, and Detlef Prescher. Statistical grammar models and lexicon acquisition. In *Linguistic Form and its Computation*. CSLI Publications, Stanford, 2001.

[Jac87] Ray Jackendoff. The status of thematic relations in linguistic theory. *Linguistic Inquiry*, 18(3):369–411, 1987.

[JH02] Leveling J. and Helbig H. A robust natural language interface for access to bibliographic databases. In Sanchez B. Callaos N., Margenstern M., editor, *Proceedings of the 6th World Multiconference on Systemics, Cybernetics and Informatics (SCI 2002)*, volume XI, page 133 to 138, Orlando, Florida, Jul. 2002. International Institute of Informatics and Systemics (IIIS).

[KF64] Henry Kucera and W. Nelson Francis. *A manual of Information to Accompany a Standard Sample of Present-Day Edited American English for Use with Digital Computers*. Providence, Linguistics Department, Brown University Press, 1964.

[KF67] H. Kucera and W.N. Francis. *Computational Analysis of Present-Day American English*. Providence, Brown University Press, 1967.

Bibliography

[KW01]   Claudia Kunze and Andreas Wagner. Anwendungsperspektiven des GermaNet, eines lexikalisch-semantischen Netzes für das Deutsche. In Bernhard Schröder Ingrid Lemberg and Angelika Storrer, editors, *Chancen und Perspektiven computergestützter Lexikographie*, volume 107 of *Lexicographica Series Maior*, pages 229–246. Niemeyer, Tübingen, Germany, 2001.

[KYCC96]   S. Keh-Yih, T. Chiang, and J. Chang. An overview of corpus-based statistics-oriented (cbso) techniques for natural language processing, August 1996.

[LBB+00]   Alessandro Lenci, Nuria Bel, Federica Busa, Nicoletta Calzolari, Elisabetta Gola, Monica Monachini, Antoine Ogonowski, Ivonne Peters, Wim Peters, Nilda Ruimy, Marta Villegas, and Antonio Zampolli. Simple: A general framework for the development of multilingual lexicons. *Journal of Lexicography*, 13(4):249–263, 2000.

[LMO96]   Stefan Langer, Petra Maier, and Jürgen Oesterle. CISLEX, an electronic dictionary for German. Its structure and a lexicographic application. In *Proceedings of COMPLEX, Budapest 1996*, 1996.

[MBF+90]   G. A. Miller, R. Beckwith, C. Fellbaum, D. Gross, and K. Miller. Five papers on WordNet. *Special Issue of the International Journal of Lexicography*, 1990.

[MC91]   G. Miller and W. Charles. Contextual correlates of semantic similarity. In *Language and Cognitive Processes*, volume 6, pages 1–28. 1991.

[ML01]   S. Müller-Landmann. Wissen über Wörter: Die Mikrostruktur als DTD. *Beiträge der GLDV-Frühjahrstagung*, 2001.

[Moo04]   Robert C. Moore. On log-likelihood-ratios and the significance of rare events. In *In Proceedings of the 2004 Conference on Empirical Methods in Natural Language Processing*, Barcelona, Spain, 2004.

[MS99]   Christopher D. Manning and Hinrich Schütze. *Foundations of Statistical Natural Language Processing*. The MIT Press, Cambridge, Massachusetts, 1999.

[MW96]   T. McEnery and A. Wilson. *Corpus Linguistics*. Edinburgh University Press, 1996.

[NL94]   Sven Naumann and Hagen Langer. *Parsing: Eine Einführung in die maschinelle Analyse natürlicher Sprache*. Leitfäden und Monographien der Informatik. B.G. Teubner, Stuttgart, 1994.

[PL05]   Sebastian Padó and Mirella Lapata. Cross-lingual bootstrapping for semantic lexicons: The case of framenet. In *Proceedings of the 20th National Conference on Artificial Intelligence* [AAA05], pages 1087–1092.

Bibliography

[PS94] C. Pollard and I. Sag. *Head-Driven Phrase Structure Grammar*. University of Chicago Press, Chicago, 1994.

[Pus95] James Pustejovsky. *The Generative Lexicon*. MIT Press,Cambridge, MA, 1995.

[QBW02] U. Quasthoff, C. Biemann, and C. Wolff. Named entity learning and verification: EM in large corpora. In *Proceedings of the Sixth Conference on Natural Language Learning (CoNNL-2002)*, San Francisco, 2002. Morgan Kaufmann Publishers.

[QHW98] U. Quasthoff, G. Heyer, and C. Wolff, editors. *Projekt Der Deutsche Wortschatz*. Linguistik und neue Medien. Deutscher Universitätsverlag, Wiesbaden, 1998.

[QRB06] U. Quasthoff, M. Richter, and C. Biemann. Corpus portal for search in monolingual corpora. In *Proceedings of LREC-06*, Genoa, Italy, 2006.

[RC98] Brian Roark and Eugene Charniak. Noun-phrase co-occurence statistics for semi-automatic semantic lexicon construction. In *COLING,-ACL,*, pages 1110–1116, 1998.

[RJ99] Ellen Riloff and Rosie Jones. Learning dictionaries for information extraction by multi-level bootstrapping. In *AAAI / IAAI*, pages 474–479, 1999.

[RS97] Ellen Riloff and Jessica Shepherd. A corpus-based approach for building semantic lexicons. In Claire Cardie and Ralph Weischedel, editors, *Proceedings of the Second Conference on Empirical Methods in Natural Language Processing*, pages 117–124. Association for Computational Linguistics, Somerset, New Jersey, 1997.

[Sch98] M. Schulz. *Eine Werkbank zur interaktiven Erstellung semantikbasierter Computerlexika*. PhD thesis, FernUniversität Hagen, Fachbereich Informatik, Hagen, Germany, Aug 1998.

[STST95] A. Schiller, S. Teufel, C. Stöckert, and C. Thiele. Vorläufige Guidelines für das Tagging deutscher Textcorpora mit STTS. *Manuscript*, 1995.

[SW95] Tony C. Smith and Ian H. Witten. Learning language using genetic algorithms. In *Learning for Natural Language Processing*, pages 132–145, 1995.

[TR02] M. Thelen and E. Riloff. A bootstrapping method for learning semantic lexicons using extraction pattern contexts, 2002.

[WW00] editor Wolfgang Wahlster. *Verbmobil: Foundations of Speech-to-Speech Translation*. Springer, Berlin, 2000.

www.ingramcontent.com/pod-product-compliance
Lightning Source LLC
Chambersburg PA
CBHW070303010526
44108CB00039B/1755